From Muslim fortress to Christian castle

D1808040

To Arthur Maass

From Muslim fortress to Christian castle

Social and cultural change in medieval Spain

Thomas F. Glick

Manchester University Press
Manchester and New York

distributed exclusively in the USA and Canada by St. Martin's Press

Copyright © Thomas F. Glick 1995

Published by Manchester University Press
Oxford Road, Manchester M13 9NR, UK
and Room 400, 175 Fifth Avenue, New York, NY 10010, USA

Distributed exclusively in the USA and Canada
by St. Martin's Press, Inc., 175 Fifth Avenue, New York, NY 10010, USA

British Library Cataloguing-in-Publication Data
A catalogue record is available from the British Library

Library of Congress Cataloging-in-Publication Data applied for
Glick, Thomas F.
From Muslim fortress to Christian castle: social and cultural change in medieval Spain / Thomas F. Glick.
p. cm.
Includes bibliographical references (p.).
ISBN 0-7190-3348-9 — ISBN 0-7190-3349-7 (pbk.)
1. Spain—History—711-1516. 2. Spain—History—711-1516—Historiography. 3. Archaeology and history—Spain. 4. Social change—Spain. 5. Land settlement—Spain. 6. Irrigation—Social aspects—Spain. I. Title.
DP99.G46 1995
946',02—dc20 95-2195
CIP

ISBN 0 7190 3348 9 *hardback*
0 7190 3349 7 *paperback*

First published 1995
99 98 97 96 95 10 9 8 7 6 5 4 3 2 1

Printed in Great Britain
by Redwood Books, Trowbridge

Contents

Figures, maps and tables

Figures

Maps

Tables

Glossary

acequia irrigation canal
Al-Andalus that part of the Iberian Peninsula under Muslim rule
alquería village (Arabic, *qarya*)
amīn inspector, particularly of a guild
Andalusi resident of, or pertaining to, Al-Andalus
apeo land survey
caliphal period from the beginning of the reign of ᶜAbd al-Raḥmān III in 912 through the fall of the Caliphate in 1009
clepsydra water clock
cortijo, cortijada large estate
despoblado deserted village
dhimmī a member of a religious minority protected under Islamic law, typically Jews and Christians
dula irrigation turn; water measure (Arabic, *dawla*)
emiral period from the establishment of the Umayyad emirate by ᶜAbd al-Raḥmān I in 756 to the end of the reign of ᶜAbd Allāh in 912
extensive archaeology the archaeology of settlements based on broad-gauged surveys, frequently guided by aerial photography, and generally with only surface collection of pottery fragments
faqīh Muslim jurisprudent, religious scholar
feudalism political system characterised by the privatisation of previously public jurisdictions, in which taxes are paid to private lords rather than into a public fisc
fitna (Arabic) civil unrest
fuero, Catalan, *fur* written law code
ḥabis, ḥabus (Andalusi Arabic) religious trust (= classical Arabic *waqf*)
ḥiṣn, pl. *ḥuṣūn* (Arabic) castle
incastellamento movement of reorganisation of countryside into castral zones
qā'id (Arabic) castellan of a *ḥiṣn*
qāḍi Islamic religious judge
qanat water filtration gallery
qarya see *alquería*
Moriscos Spanish Muslims of the sixteenth century, forcibly converted to Christianity
Mozarabs Christians living in Al-Andalus under Muslim rule
Mudéjars Muslims living under Christian rule up to 1492
Muwallads members of Hispano-Roman population converted to Islam after the Muslim conquest of Spain
noria hydraulic wheel for lifting water, drawn by a mule, ox, or donkey
qawm tribal segment, clan
rābiṭa, ribāṭ (Arabic) frontier religious settlement of warriors
Repartimiento (Castilian), *Repartiment* (Catalan) apportionment to Christian settlers of land and houses captured or confiscated from Muslims

shādūf swape, balanced bucket used to lift water
Sharīʿa Islamic law
Sharq al-Andalus the east of Al-Andalus, including Valencia and the Balearic Islands
taha (Arabic) administrative district
tahulla measure of value of agricultural land
tapia rammed-earth technique for building walls
tributary state one whose governmental system is based on taxes paid into a public fisc, as contrasted with a feudal society in which rents are paid to private lords
Upper March frontier region in what is now Aragón and Alto Aragón eastward to Lleida
veedor Castilianisation of Arabic *amīn* (v.)

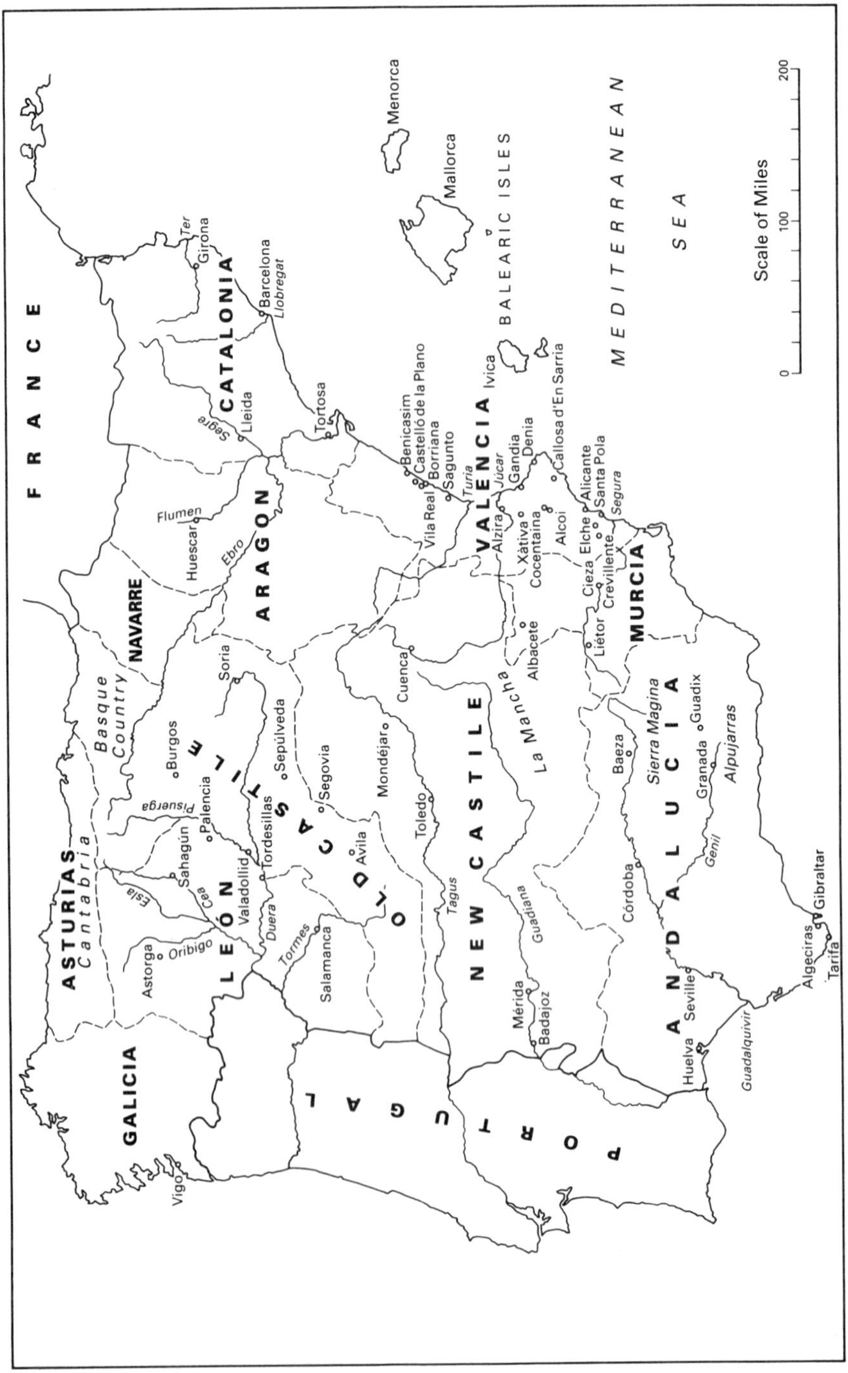

Map 1 *Spain in the sixteenth century*

Preface

Three key transitions

The energies unleashed by Pierre Toubert's conception of *incastellamento* (which was an invitation to look at castles as a social phenomenon) and the famous Rome meeting of 1978 had two spectacular results in Spanish medieval studies. The first was the overthrow of the received view of feudalism. Medieval Spain had always been characterised as 'incompletely feudalised', with considerable freedom available to peasants willing to move to the frontier. What historians had failed to notice was that such peasants quickly lost their freedom as the frontier advanced and the old frontier receded into the hinterland. Spanish historians now regard Spain as one of the *most* feudalised societies of medieval Europe.

A second result was to invigorate medieval archaeology in order to delineate the process of *incastellamento*, on the feudal Christian side of the border but, more significantly, on the non-feudal Muslim side. The result has been a data explosion without precedent in medieval Spanish historiography which historians outside Spain have scarcely begun to digest. Previously we knew next to nothing about rural settlement and social organisation in Al-Andalus, that is, Islamic Spain. In Spain, these energies were diffused by French scholars working under the aegis of the Casa de Velázquez (Pierre Guichard, André Bazzana, Patrice Cressier), and the method then spread to Catalan and Spanish historians working in the study areas represented.

The 'archaeological revolution' had not yet begun in earnest when I wrote *Islamic and Christian Spain in the Early Middle Ages* (Princeton, 1979). When I revised that book for a Spanish publisher, I strove to integrate as many of the findings of the archaeologists that I deemed relevant.[1] But the book was designed to represent a distinctly historical 'register' and I came to feel that the appearance of an archaeological register required a different structure.

The result is the present volume, which is meant to be read at

1 *Cristianos y musulmanes en la España medieval* (Madrid, Alianza, 1991).

two different levels simultaneously. The first 'track' – the historical – is about socially-induced processes of landscape change in the Iberian peninsula across two cultural transitions and a sociopolitical one. Chapters 1 and 2 describe the transition from the Roman to the Islamic landscape of the early Middle Ages according to the findings of archaeologists generating new hypotheses about this process. Chapter 3 takes a periodisation derived from the material records of the transition and broadens it to include social and intellectual phenomena that give greater body to a 'Paleoandalusi' period extending from the Arab/Berber conquest to the beginning of the tenth century. Chapter 4 describes in greater detail a significant element of the first transition – irrigation. Chapter 5 is about the reorganisation of the landscape that took place in Christian Spain in the eleventh century, under the process of *incastellamento.* Chapter 6 recounts the second cultural transition, that of the lower Middle Ages, using the *Libros de Repartimiento/Llibres de Repartiment* as sources for the evaluation of landscape change. Chapter 7 shows how, in that transition, the processes informing landscape change can be extrapolated to other areas of experience.

The second track is historiographical and methodological. In this context, chapter 2 is about how a 'universalising' hypothesis of landscape change and organisation of rural society was used to force debate on a multiplicity of related issues. Chapter 3 is a deliberate attempt to use an archaeologically-generated periodisation to substantiate a view of Andalusi history in which the orientalising reign of ᶜAbd al-Raḥmān II had been viewed as a turning-point in the history of Al-Andalus, but on substantially narrower grounds than those portrayed here. Moreover, an older universalising hypothesis, Richard Bulliet's retrodictive account of the processes of religious and 'social' conversion, is shown to be supported by the new synthesis which I present. In Chapter 4, on irrigation, I take a narrow set of hypotheses emanating from the new model of rural social organisation in Al-Andalus and broaden them into a general hypothesis on the nature of local control in the allocation and use of water. Chapter 5 is an excursus on the historiography of the 'transition to feudalism' in which a number of different schools have used the same key terms – transition and feudalism – to explain complementary, but very different, processes. Here I show that the cross-cutting solidarities that the political organisation of feudalism encouraged had a number of paradoxical effects, particularly in the allocation and use of water.

One of this book's objectives, therefore, is to guide the reader through the maze of archaeological results, virtually unnoticed in recent English literature.[2] For example, in a volume devoted to the eighth century (that is, to the Islamic conquest of Visigothic Spain), Roger Collins avoids introducing the archaeological debate completely. This leaves him with little to talk about except the dim reflection of a failed Church and the dissolution of Toledo as the religious as well as the civic capital.[3] But this is a strange window indeed on the century which witnessed the birth of a spectacular new centre of civilisation.

An otherwise excellent survey of 'Moorish Spain' by Richard Fletcher likewise shows no awareness of the historiographical and archaeological revolution that has taken place, nor mentions a single one of its architects.[4] The book represents the standard scholarly view of Islamic Spain as it was in 1975, before the publication of Pierre Guichard's study recasting Andalusi society in a tribal context.[5] (Guichard's subsequent career has been mainly devoted to the archaeological revolution.)

It appears that an entire generation of English-speaking historians of medieval Spain has lost control of the research front at the same time. Peter Linehan, in a book ostensibly about the recent historiography of medieval Spain, completely misses the point that the historiographical revolution was closely linked to the archaeological one, based on the practice of 'extensive archaeology', which concentrates on surface features rather than on traditional stratigraphic methodology and which had produced such striking, even revolutionary, results, in the areas surveyed (Valencia, Mallorca, Almería).

Linehan's assessment of the new doctrine of Spanish feudalism is wide of the mark. First, he takes the story only to 1980, and therefore does not perceive Toubert's influence. Second, by 'Spanish', he means Castilian, and the new Castilian historiography of feudalism he deems to have been victimised by 'academic Marxists'. He regards it as ironic that Léon–Castile is now regarded as a fully feudalised society just when, in the rest of Europe, 'the twin dogmas of normative feudalism and the fundamental uniformity of

2 And surprisingly unnoticed in Spanish historiography, even that of Al-Andalus; see Manuel Acién Almansa's plaint, 'Arqueología medieval en Andalucía', in *Coloquio Hispano-Italiano de Arqueología Medieval* (Granada, Patronato de la Alhambra, 1992), pp. 27–33, on p. 32.

3 Roger Collins, *The Arab Conquest of Spain, 710–797* (Oxford, Basil Blackwell, 1989).

4 Richard Fletcher, *Moorish Spain* (New York, Henry Holt, 1992). The studied archaism of 'Moorish' is an indication of his approach to the subject.

5 Pierre Guichard, *Al-Andalus: Estructura antropológica de una sociedad islámica en occidente* (Barcelona, Barral, 1976).

feudal institutions' have been abandoned.[6] Linehan has a problem perceiving anything beyond the borders of Castile as 'Spanish'. Moreover, his continual derogation of those who reject 'national' history suggests an animus against, in particular, Catalonia, whose title to nationhood is, by inference at least, rejected. But Guichard is no Catalan; and the fact that he, by stressing the non-feudal nature of *incastellamento* in Al-Andalus, provides the analytical basis for a reassessment of feudalism in Catalonia and Valencia, is lost on Linehan, who mentions neither Guichard nor *incastellamento.* There is ample irony here because Linehan, in a perceptive, positive review of Guichard's monumental book on tribal structures in Al-Andalus, noted that Spanish medievalists would have to respond to Guichard's analysis.[7] Here he had the chance, but seems unaware of Guichard's role in forging the very revolution he decries. Instead, he supplies banalities on the 'European union' impulse of the new breed of Spanish medievalists who desire only to abandon Spain's particularism.

This revolution took place at the same time as the old 'liberal' doctrine of feudalism associated with Claudio Sánchez-Albornoz was being overthrown in the 1970s and 1980s, and therefore the reassessment of feudalism in Catalonia and Valencia was very closely tied to Toubert and *incastellamento* and was expressed in an explanatory style whose centrepiece was the contrast between the function of castles in feudal Catalonia/Valencia on the one hand, and Al-Andalus on the other. Among Castilian medievalists, however, the reassessment of feudalism took place in isolation from the *incastellamento* debate, and therefore the break with the 'liberal' model has been conceptually more narrowly focused on the issue of dependency, rather than on overall social organisation.[8]

The debate that Guichard's work stimulated has entirely changed the focus of early medieval Andalusí history, from dynastic history to rural settlement patterns: as drastic or polar a change as one could possibly conceive in historical research. Archaeological debate turns on two issues: first, the nature of rural settlement at the time of the Muslim conquest; and second, the subsequent reorganisation of rural settlement by tribally organised groups (Arabs and Berbers). Nor are these issues parochial: the role of castles and

6 Peter Linehan, *History and Historians of Medieval Spain* (Oxford, Clarendon Press, 1993), pp. 191–200, esp. p. 197.

7 *Social History,* 3 (1978), 377–9.

8 On the recent historiography of Iberian feudalism, see Reyna Pastor, 'Sobre la constitución y consolidación del sistema feudal castellano–leonés de los siglos XI–XII', *Estudi General,* 5–6 (1985–86), 199–210, on pp. 200–1; and Carlos Estepa Díaz, 'In Memoriam: Abilio Barbero de Aguilera (1931–1990)', *En la España Medieval,* 14 (1991), 11–17.

the nature of social and political structures in rural Al-Andalus are direct offshoots of the discussion of the significance of *incastellamento*.

These dramatic results represent the completion of a long historiographical cycle. Medieval archaeology before 1970, approximately, had been a politically reactionary field in Spain, focused mainly on the Visigoths even before the Franco period, in an attempt to justify the supposed continuity of the Visigothic kingdom of Toledo with the Asturian kingdom and to legitimise the idea of 'Reconquest', in other words to press a 'Pan-Hispanic' historiographical and ideological programme. This line of research, which in the 1930s had attended to the 'Germanisation of Spain' and sought to bring Spaniards within the compass of the 'German race', was vigorously adopted by the Franco regime, which had no ideology and cobbled one together out of 'eternal Spain', Reccared, Covadonga, the 'Reconquest', and the Catholic Kings. The same regime that 'purified' the best Islamic archaeologist, Leopoldo Torres Balbás, was loath to look too closely at Islamic culture nor even to pursue much of a programme in 'Christian archaeology', which focused on castles and churches and whose meagre material remains pale in comparison to the richness of Islamic civilisation.[9]

Historiographical debates and polemics, many of which are discussed in this book, though frequently self-serving, have the positive result of forcing a higher level of theorisation. Although seemingly over-personalised, these debates betray the thinking of very different professional cultures as well as the struggle that a new field, medieval archaeology, has had in legitimising its place in the post-Franco academic structure. This too helps explain the fury of the polemics that the Guichard thesis has unleashed. The three basic issues he has raised – tribalism, the nature of rural social organisation, and the extent of Berber settlement – have all been received with hostility by Arabists who, because of the philological and textual nature of their academic culture, have been weak in generating social hypotheses.[10] The virtual silence of Spanish Arabists with regard to Richard Bulliet's hypothesis on the nature and rate

9 For the political history of Spanish medieval archaeology, see Lauro Olmo Enciso, 'Ideología y arqueología: los estudios sobre el período visigodo en la primera mitad del siglo XX', in Javier Arce and Ricardo Olmo, eds., *Historiografía de la arqueología y de la historia antigua en España (siglos XVIII–XX)* (Madrid, Ministerio de Cultura, 1991), pp. 157–60; and Vicente Salvatierra Cuenca, *Cien años de arqueología medieval. Perspectivas desde la periferia: Jaén* (Granada, Universidad, 1990), pp. 72, 76–7. On the *depuración* of Torres Balbás, see Carlos Vílches Vílches, *La Alhambra de Leopoldo Torres Balbás* (Granada, Comares, 1988), p. 515.

10 On the incapacity of Arabists to conceptualise historical problems, see the trenchant remarks of Miquel Barceló, 'Quina arqueologia per al-Andalus?', in *Coloquio Hispano-Italiano de Arqueología Medieval* (Granada, Patronato de la Alhambra, 1992), pp. 243–52, on p. 247.

of conversion to Islam in Al-Andalus is a case in point.[11] Berberisation has been a focal point of the polemic, in some cases for ideological reasons (e.g., the difficulty Arab historians and many Western Arabists have in authenticating minority cultures in the Arab world), in the rest because of the radically divergent orientations of different professional cultures. Because of the overly rigid disciplinary boundaries of Spanish Arabism, Berber culture, both its material remains and that part reflected in standard Arabic historical sources, falls outside the professional control of Arabists. Hence, the vengeance with which the new archaeologists have picked up this particular banner. Moreover, the style of archaeological explanation, which is to promote very broad hypotheses in order to push research in specific directions, seems excessively categorical to historians unfamiliar with this explanatory tradition.[12]

Over the past decade, I have been critical of specific points of Guichard's thesis of rural social organisation. We wrangled in footnotes over the meaning of specific cultural markers I professed to have found in Valencian irrigation systems.[13] I questioned whether it was logical that Berbers should leave traces of their settlement patterns, but no other cultural markers;[14] I wondered whether Guichard's characterisation of the *alquería* as a free commune wasn't a romanticisation;[15] I questioned the notion that seemed implicit in his concept of small-scale *alquería* irrigation that a poor settlement with rudimentary material conditions requires a rudimentary irrigation system;[16] that concentration on tribal settlement leaves out the Muwallad/s (the indigenous majority of Hispano-Romans

11 Richard Bulliet, *Conversion to Islam in the Medieval Period: An Essay in Quantitative History* (Cambridge, MA, Harvard University Press, 1979), Ch. 10. See my discussion of Bulliet's hypothesis in *Islamic and Christian Spain in the Early Middle Ages* (Princeton, Princeton University Press, 1979), pp. 33–5. A thoughtful, but neutral response, by a Spanish Arabist to Bulliet, is María J. Viguera Molíns, 'Sobre Mozárabes', in *Proyección histórica de España en sus tres culturas: Castilla y León, América y el Mediterráneo*, vol. III, *Arabe, hebreo e historia de la medicina* (Valladolid, Junta de León y Castilla, 1993), pp. 205–16, on p. 215.

12 See, in this regard, James A. Bell, 'Universalisation in Archaeological Explanation', in Lester Embree, ed., *Metaarcheology: Reflections by Archaeologists and Philosophers* (Dordrecht, Kluwer, 1992), pp. 143–63.

13 Guichard, *Al-Andalus*, pp. 304–5; Thomas F. Glick, 'Las técnicas hidráulicas antes y después de la conquista', in *En torno al 750 aniversario: Antecedentes y consecuencias de la conquista de Valencia*, 2 vols. (Valencia, Consell Valencià de Cultura, 1989), p. 59.

14 Thomas F. Glick, 'L'alta edat mitjana', in *Història del País Valencià* (Valencia, Tres i Quatre, 1992), pp. 57–82, on p. 62.

15 *Ibid.*, p. 66.

16 Thomas F. Glick, 'Berbers in Valencia: the Case of Irrigation', in *Medieval Spain and the Western Mediterranean: Essays in Honor of Robert I. Burns, S. J.* (Leiden, E. J. Brill, 1995).

converted to Islam), and so forth. But finally I have become convinced of the correctness of his system. First, it is logical. Critiques of it have been scatter-shot, have not undermined its coherence, and offer no alternative explanation that can be supported by the evidence. Second, new evidence turned up by persons not directly in Guichard's orbit, and particularly by a younger generation of Spanish historians who know Arabic, tends to support his theses. Third, Josep Torró's synthesis of Guichard's approach to rural settlement showed me how it could be converted into a series of hypotheses testable by historiographical means.[17]

Archaeology presents a series of challenges to medieval history, first and foremost being the need to theorise cultural change or, rather, to conceptualise it at a finer level of definition than has heretofore been the case. This volume is a contribution to that complex task. As Miquel Barceló has observed, history and archaeology yield two different archives, 'difficultly complementary'. He goes so far as to say that the two archives ('registers') do not so much illuminate each other as pose a quite different order of problems. The world revealed by archaeology is a rural world, that inhabited by peasants and one (in the case of Islamic Spain) not at all revealed in written texts.[18] David Hall has remarked, in quite a different context, that 'the culture of the European peasant may be likened to a river full of debris . . . a muddied, multilayered process by which culture was transmitted, one that functioned to preserve and pass along many bits and pieces of past systems of belief'[19] and (I might add) social organisation. I will not here attempt to resolve the epistemological quandary suggested by Barceló; rather my objective is the instrumentalist one of suggesting the utility of the archaeological archive – the register of 'debris' – to historical explanation.

Al-Andalus (in Barceló's phrase) is a 'palimpsest, difficult to read'.[20] That reading requires an adequate level of cultural theorisation. My original interest in culture contact in medieval Spain was stimulated in the early 1960s, by reading classic works by A. L. Kroeber and Américo Castro. It seemed to me then, as it does now, that the Kroeberian perspective on culture change supplied a simpler and clearer explanation for what Castro was attempting to

17 Josep Torró, *Poblament i espai rural: Transformacions històriques* (Valencia, Institució Valenciana d'Estudis i Investigació, 1990).
18 Miquel Barceló, 'Historia y arqueología', *Al-Qantara*, 13 (1992), 457–62, on pp. 458–9.
19 David D. Hall, *Worlds of Wonder, Days of Judgment* (New York, Knopf, 1989), p. 11.
20 Barceló, 'Quina arqueologia per al-Andalus', p. 250.

explain. I read Levi-Strauss's *Structural Anthropology* at the same time, found Kroeber more accessible and convincing, and developed a distrust of 'deep structures', either in society or in culture. I am much more comfortable with Giddens's dynamic model of cultural change (which he calls 'structurationist') that stresses the recursivity of social and cultural phenomena. By recursivity, I mean that societies and cultures constantly change under the influence of daily experience, which interacts recursively with received structures and alters them.[21] Castro was deaf to the kind of mute evidence that makes up the archaeological palimpsest, which is, no doubt, why archaeologists see nothing that attracts them in his work. But the dismissal of Castro as nothing more than an idealist 'metahistorian' by the new archaeologists amounts to a failure to sort out what was valid in his work from the weak philosophical underpinnings that Castro infelicitously borrowed from German idealism.[22] Castro can and should be read in a 'structurationist' sense: that lived experience (his '*vividura*') can alter social and cultural norms and practices to the point where new ethnic entities are created and defined: thus Visigoths were not 'Spaniards', nor did the Christians of the early Middle Ages have any authentic connection with the Visigothic tradition.[23]

Inasmuch as the history of Al-Andalus and its civilisation turns on the settlement there of Arab and Berber tribal groups who dominated the indigenous 'Hispano-Roman' population, diffusion and acculturation are an important part of the story. Yet the archaeological literature is singularly undertheorised in this regard: the archaeologists have a rather carefully developed theory (or theories) of social structure and action (segmentary tribal society; tributary governmental organisation), but no theory of culture to speak of. Barceló goes so far as to oppose using the term 'culture' itself because, in his view, it has been 'manipulated and mystified'.[24] From the perspective of Spanish historiography, with its backwash of Pan-Hispanic idealism or current ideologised debates about the

21 I go into somewhat greater detail on this view of social change, derived from Anthony Giddens, in *Cristianos y musulmanes en la España medieval*, pp. 20–1.

22 Vicente Salvatierra Cuenca, *Cien años de arqueología medieval. Perspectivas desde la periferia*: Jaén (Granada, Universidad, 1990), p. 76, is a standard dismissal of both Castro and Sánchez-Albornoz as idealist 'metahistorians'. In Valencian historiography, the misguided notion that Castro and Sánchez-Albornoz had comparable ideological suppositions seems to have originated with Joan Fuster. See Ferran García Oliver, *Terra de feudals* (Valencia, Edicions Alfons el Magnànim, 1991), p. 20 n. 4.

23 Américo Castro, *The Spaniards* (Berkeley, University of California Press, 1971), p. 183.

24 Barceló, 'Quina arqueologia per al-Andalus?', p. 244.

'essence' of Valencian 'culture', there is ample justification in his sentiment, but not in his conclusion. A concurrent, but also outmoded, Marxist critique of cultural anthropology which objects to any notion that culture may in some way be independent of economic infrastructure has no doubt also influenced this draconian assessment. But by 'theory of culture' I mean something much more concrete than the current vogue of chic anthropologism whereby 'cultural' historians explain everything from diet to vernacular architecture in terms of *mentalité*. Barceló's objection to this kind of reductionism is right on the mark.[25]

Much of this volume deals with settlement patterns and how peoples of differing cultures and social organisation organised space. In part this is simply a reflection of the kind of phenomena that the new archaeologists have studied, and clearly the subject of extensive archaeology is settlement. In looking for documental reflections of the late medieval transition, that which begins in the thirteenth century, there is a ready-made body of literature, the great books of land division (*Repartimientos*), which by virtue of their rationale pertain to settlement patterns and little else. However, there is also a good *a posteriori* reason for concentrating on settlement and organisation of space: in comparative study one looks for coherent sets of phenomena, particularly those that package together a multiplicity of cultural elements. With many elements the pattern which integrates them into a whole itself becomes a valid cultural marker.[26]

Irrigation, hydraulic technology, and water management are an integral part of the processes of cultural change described in this book, a sub-set of the more inclusive package of settlement and land use. Water allocation and the diffusion of hydraulic techniques from east to west are responses to specific geographical problems by peoples with different social and economic organisations. But the diffusion and implantation of both techniques and their specific application, including institutional arrangements governing water allocation and distribution, involve *mental* as well as economic and social processes; that is why a theory of culture cannot be dispensed with. Irrigation was the linchpin of much of Islamic and Christian agriculture. But the two cultures went about its management in quite different ways and both irrigation and the use of water power for milling were integrated differently into each society. Therefore water use becomes a highly significant social and

25 *Ibid.*, p. 249.

26 Cf. my remarks in 'New Perspectives on the *Ḥisba* and its Hispanic Derivatives', *Al-Qanṭara*, 13 (1992), 465–89, on p. 475.

cultural marker. More than thirty years of conversations and interchange with Arthur Maass on water and its cultural expressions have radically shaped my approach to all problems of cultural and historical interpretation. The simple lesson that institutional structures must be explained in terms of their function I learned from him. Moreover, his specific approach to irrigation institutions – that regulations and customary practices embody and encode social values and objectives – furnishes a general methodology for comparative study.

Originally I had hoped to be able to define with some precision the transition from Roman Spain to Al-Andalus. Although I have supplied as many hints as I can to the dimensions of that shift, neither the archaeological nor the historical record are complete enough to permit the implementation of such a programme. Classical archaeologists in Spain have not studied Roman settlement patterns to anywhere the extent necessary to be able to describe that critical transition. Therefore I have had to redefine the first transition as that leading to *incastellamento*, more like the transition from the 'Dark' to the 'Middle' Ages than from antiquity to post-antiquity. This entails a certain lack of phase between Al-Andalus and Christian Spain because it is possible (to a point) to describe the process of settlement in Islamic Spain – Al-Andalus – from the eighth century on, on the basis of archaeological evidence, while in Christian Spain, with little archaeological research, I am content to pick up the story in the ninth century.

The second transition considered is the transition to feudalism in Christian Spain, a set-piece in western historiography. Here it is introduced to show the similarities and differences between social transitions on the Christian and Muslim side of the border. Moreover, the feudal transition is important relative to the historiographical concerns of this book: the doctrine of a highly feudalised Christian Spain in part grew out of a vision of a non-feudalised *incastellamento* in Al-Andalus.

The third transition, from Islamic to late medieval Christian Spain, while conceptually clearer, also entails a considerable chronological spread. I take the beginning of that transition from the conquest of Toledo in 1085, when the ingestion of large numbers of Muslims first became an issue and when the landscape to be transformed was heavily 'orientalised'. One might conclude the discussion with the great conquests of the thirteenth century, the *Repartimientos*, and subsequent adaptations of settlement. But I have chosen to include under this rubric the Granadan transition of the sixteenth century as well: in spite of the fact that this period extends from the conquests of 1492 right through the expulsion of the Moriscos of 1611, the documents generated reveal the same modal change of settlement that has interested me in earlier periods, but in greater detail.

Keeping abreast of the rapidly changing research front in archaeology has meant relying on drafts and on information and impressions received orally about work in progress. Archaeologists will frequently give you their sense of the significance of a particular site, impressions that they would not yet put into a printed report. They form impressions that are difficult to document, a sense of the rhythm of change. In this respect, important contacts have included Manuel Acién Almansa, Rafael Azuar Riuz, Miquel Barceló, Patrice Cressier, Manuel Espinar Moreno, Sonia Gutiérrez Lloret, Helena Kirchner, Joan Mateu Bellés, Carmen Navarro, Lauro Olmo Enciso, Angel Poveda Sánchez, Josep Torró, and Juan Zozaya.

I am grateful to Antonio Gómez Becerra for permission to reproduce Maps 2 and 3, and to Sonia Gutiérrez Lloret for Maps 5 and 6. Map 7 is redrawn from Philippe Araguas, 'Le réseau castral en Catalogne vers 1350', *Castrum 3*, pp. 114–15.

PART ONE

The early Middle Ages

1

The late Roman landscape

In order to assess subsequent changes in the rural landscape, we must have at least a minimal picture of the Roman landscape before us. In the literature of medieval settlement it is typically assumed that the Christian *aldea* or Muslim *alquería* was simply the heir of the Roman villa. But it is virtually impossible to substantiate such claims because no one knows what the late Roman villa, much less that of the Dark Ages, was like, or how – exactly – a private agricultural establishment with a single wealthy owner evolved into a peasant village.

After the economic crisis of the third century, the villa enjoyed a renaissance in the fourth, during which the great agricultural domains reached their apogee. At this time, agriculturally oriented *villae rusticae* were established in places that had been only lightly Romanised previously: in Galicia, for example, where there are an abundance of place-names based on *vicus, fundus,* or villa, or in Asturias where rural *villae* became centres of settlement as well as hearths of Romanisation.[1] In the older, Romanised areas like Betica (where, in the olive- and wheat-growing region between Seville and Córdoba, density of settlement was comparable to that of Roman Italy), Tarraconensis (with 52 per cent of Roman agricultural establishments in the fourth century) and Lusitania, large numbers of *villae* were built or enlarged in the course of the fourth century, just as Roman cities embarked upon a long decline. Thus in Gorges's view, the decadence of fourth-century cities was offset to a degree by the reconstitution in the countryside of these large rural nuclei which had some urban functions (they were nodes for local and regional commercial exchange, for example).[2]

Historian Javier Arce, on the other hand, dissents from the

1 Jean-Gerard Gorges, *Les villas Hispano-Romaines* (Paris, E. de Broccard, 1979), pp. 48, 51.

2 *Ibid.*, pp. 55, 82. Kevin Greene, *The Archeology of the Roman Economy* (Berkeley, University of California Press, 1986), p. 110. However, the latter author warns (p. 114) that density of sites in Betica and Tarraconensis may be genuine or just an epiphenomenon of modern survey methods.

common view of urban decline presented by archaeologists and is insistent that the Roman city was transformed in late antiquity, but did not disappear. There was an urban life, 'and not one that was ruinous, destroyed, forgotten, as can appear in archaeologists' reports . . . . Of necessity, the pessimistic, disastrous and desolate vision of the archaeologists must be wrong.'[3] But archaeologists now certainly take a more nuanced position, that many Roman towns survived with their political control practically reduced to their immediate hinterlands, while others disappeared (for example, Begastri, a town mentioned in the 'Pact of Theodomir').[4]

The difference between villa and *vicus* is not clear in late Roman (or Visigothic) sources. For Varro, a *vicus* was just a group of houses. The term was typically applied, however, to Romanised or semi-Romanised native settlements, in contradistinction to Roman *villae.* (Thus Vigo, in Galicia, derives its name from *vicus.*) In Roman inscriptions, *vicus* is usually followed by a tribal name in the genitive plural or some other element indicative of locale.[5] A *fundus* was a group of *villae*, possibly with a mean extension of some 1,000–1,500 ha, while the *villa* proper rarely exceeded 2 ha. The same property could extend through various *fundi.* Finally the *pagus* was a Roman administrative unit for the purpose of tax collection, and was found in precisely those areas where there was a large agrarian population to be taxed.[6] These *fundi* tended to be of modest size in the northern, wetter regions of Hispania, typically 25–50 ha in Galicia or Catalonia, while in Lusitania there were huge *latifundia*, estates of 5,000–10,000 ha not being rare.[7]

Villae were typically built on small hills or hillsides overlooking their farmland. They were typically clustered around cities or along Roman roads, as in the case, for example, of the periurban concentration of *villae* in a 20-kilometre radius around Salmantica –

3 Javier Arce, 'La transformación de Hispania en época tardorromana: Paisaje urbano, paisaje rural', in *De la Antigüedad al Medioevo, siglos IV–VIII. III Congreso de Estudios Medievales* (Madrid, Fundación Sánchez-Albornoz, 1993), pp. 225–49, on pp. 243–4. Leonard A. Curchin, *The Local Magistrates of Roman Spain* (Toronto, University of Toronto Press, 1990), p. 120, also argues for an economic revival in fourth-century Roman Spain.

4 Sonia Gutiérrez Lloret, 'De la civitas a la madīna: Destrucción y formación de la ciudad en el sureste de Al-Andalus. El debate arqueológico', *IV CAME*, I, 13–35.

5 Leonard A. Curchin, '*Vici* and *Pagi* in Roman Spain', *Revue des Etudes Anciennes*, 87 (1985), 327–43, on pp. 329, 335.

6 Curchin, '*Vici* and *Pagi*', p. 342; E. Cerrillo Martín de Cáceres and José María Fernández Corrales, 'Contribución al estudio del asentamiento romano en Extremadura. Análisis espacial aplicado al S. de Trujillo', *Norba*, 1 (1980), 157–75, on p. 159.

7 Gorges, *Villas Hispano-Romaines*, pp. 94, 98.

Salamanca – mainly on the banks of Tormos, or the seventy *villae* within 30–40 kilometres of Mérida, typically located to the south of the city in alluvial bottoms of small valleys with access to the highway to Seville.[8] Late Roman *villae* fortified with a tower – *turris* – would appear to be the origin of numerous medieval villages in Torre- (e.g., Torrejón, Torrequemada – a ruined fort, etc.). How such places evolved from fortifications into rural villages is not known.[9] Zozaya notes dense concentrations of *villae* along roads which followed the courses of rivers, as in Soria, with heights dominated by *turres castellae.* The *turres* guarded the fields of the *dominus,* and the concentration of *villae* make understandable the demographic basis of small private armies.[10]

The invasions of the fifth century heightened the character of refuge that many fourth-century *villae* – as poles of attraction for those fleeing the cities – had, and favoured those with defensively advantageous emplacements. 'In these troubled times, many of the old, unfortified vici were abandoned . . . and the villas with their outlying dependencies became the basis for the villages of the Middle Ages.'[11] Gautier-Dalché supposed the general structure of habitat south of the Pyrenees on the eve of the Muslim invasion wasn't much different from those of Gaul and Italy – the network of *villae* and *vici* was therefore 'probably in great part intact'. Archaeologists now doubt this optimistic conclusion.[12]

In the Alicante region, very few *villae* survived into the sixth or seventh centuries, ruralisation having made its mark from the fifth century in new hilltop sites such as El Castellar (Alcoi) and Pic Negre (Cocentaina). Cities were in full regression, Dianum (Denia) from the fifth century. Portus Illicitanus (Santa Pola, the port of Roman Elche) had disappeared by the end of the sixth century, a sign of the failure of the Roman commercial network in the western

8 *Ibid.*, p. 87, and Jean-Gérard, Gorge, 'Implantations rurales et réseau routier en zone en zone meritaine: Convergences et divergences', *Caesarodunum,* 18 (1983), 413–24; and 'Prospections archéologiques autour d'Emerita Augusta: Soixante-dix sites ruraux en quête de signification', *Revue des Etudes Anciennes,* 88 (1986), 215–36.

9 José M. Fernández Corrales, 'El asentamiento rural romano en torno a los cursos alto y medio del Salor: su marco geográfico y distribución', *Norba,* 4 (1983), 207–21, on p. 209.

10 Juan Zozaya, personal communication, 22 August 1980; Javier Arce, *El último siglo de la España romana (284–409)* (Madrid, Alianza, 1982), pp. 76–7.

11 Gorges, *Villas Hispano-Romaines,* p. 151 and Leonard A. Curchin, '*Vici* and *Pagi*', p. 336.

12 Jean Gautier-Dalché, 'Châteaux et peuplements dans la Péninsule Ibérique (Xe–XIIIe siècles)', in *Châteaux et peuplements en Europe Occidentale du Xe au XVIIIe siècle* (Auch, Centre Cultural de l'Abbaye de Flaran, 1980), pp. 93–107, on p. 95.

Mediterranean when imports of North African and Eastern fine pottery wares first declined and then disappeared throughout the area.[13] When trade atrophied, so did the Roman roads, a phenomenon aggravated by the dislocation of population from the lowlands to more secure settlements higher up. The Roman road from Orihuela to Murcia had run through the alluvial plain in a straight line. Then, as population moved into the foothills, the old road was abandoned in favour of a tortuous one. The present-day Alicante–Murcia highway still reflects that settlement shift, running along the edge of the sierra.[14]

A related dislocation of population in late Roman times involved small groups escaping from the control of masters of *latifundia* and establishing themselves in ecologically marginal areas in fertile swampy regions of the Mediterranean coast, where they lived on subsistence agriculture together with the kind of alimentary gathering (e.g., shellfish) that residence in such areas makes possible.[15]

Continuity of settlement is a thorny issue. According to Gorges, some late Roman *villae* did survive: La Cocosa (Badajoz) until the Arab conquest; Navatejera until the Middle Ages; the Roman *villa Almenar* became a Visigothic one and survived as a *cortijo*.[16] In the Aljarafe region of Seville sites of Roman *villae* can be shown to have lasted into the Middle Ages, giving rise to *alquerías* or *cortijadas* as successor settlements. These sites also give ample evidence of the reutilisation of Roman building materials (bricks and roofing tiles) in the early Islamic period.[17]

In Catalonia the many place-names with the affix Vila- or Villa- are frequently taken as indicating continuity of occupation. But Keay notes that very few sites with such names documented in the ninth and tenth centuries correspond with known late Roman sites. Banks's view is that places so named that disappeared in the early Middle Ages may well have been Roman *villae*, but thriving medieval

13 Chris Wickham, 'L'Italia e l'alto medioevo', *Archeologia Medievale*, 15 (1988), 105–25, on pp. 110–11.

14 Sonia Gutiérrez Lloret, 'El poblamiento tardorromano en Alicante a través de los testimonos materiales: Estado de la cuestión y perspectivas', *Antigüedad y cristianismo Arte y poblamiento en el SE peninsular durante los últimos siglos de civilización romana* (Murcia), 5 (1988), 323–37; Antonio González Blanco, 'La población del sureste durante los siglos oscuros (IV–X)', *ibid.*, pp. 11–27, on pp. 19–20.

15 Sonia Gutiérrez Lloret, 'El origin de la huerta de Orihuela entre los siglos VII y XI: Una propuesta arqueológica sobre explotación de las zonas húmedas del Bajo Segura', *Arbor*, 151 (1995), 65–93, on pp. 82–3.

16 Gorges, *Villas Hispano-Romaines*, p. 57.

17 José Luis Escacena Carrasco, 'Yacimientos arqueológicos de la época medieval en el flanco oriental del Aljarafe', *II CAME*, II, 579–87.

villas or *vilas* must be reckoned of medieval foundation, unless complementary evidence can be found. The late Roman settlements of the plain of Barcelona either disappeared or lost their function as nucleated settlements, with consequent dispersion of population, except in those cases where a church associated with a villa survived to serve as a focus for nucleation. Medieval Catalans recognised the ruins of ancient *villae*, which appear in early medieval documentation as *parietes antiquas* or *delgades* (old or thin walls). But for Banks, although 'certain features of the rural settlement pattern of classical origin survived until the early seventh century . . . it is impossible to point to any definite example where there was direct continuity of settlement from Antiquity to the Middle Ages'.[18] In El Maresme, Catalonia, on the other hand, of 268 villa sites studied by Prevosti, sixty-four have a Romanesque or pre-Romanesque *masía* on top of them.[19]

Map 2 depicts Roman settlements on the coast of Granada. This was a site of the fish-preserving industry which survived until it was extinguished in the late fourth or early fifth century, a victim of the disruption of the Roman commercial network in the western Mediterranean, as well as of agriculture and mining. Mainly *villae* on the coastal plain disappeared at this time, although ceramic remains of the fifth and perhaps early sixth centuries attest to some continuity of commercial activity. The same period witnessed the displacement of population – surely dependent cultivators – from the coast to mountains of the interior, a common phenomenon throughout the western Mediterranean. Map 3, the same region in the high Middle Ages, shows the result of this movement, as emiral settlements (*alquerías*), protected by fortified castles (*ḥuṣūn*), represent in great part the successors of late Hispano-Roman settlements.[20]

Settlements of the Visigothic period are portrayed as continuations of late Roman settlements, with no sharp differentiation: Roman settlements were gradually incorporated into the Visigothic political and economic structure. The *Lex Visigothorum* mentions four kinds of settlement types: *civitas, castellum, vicum,* and villa. Goths appear to have settled in *vici*, whose residents – *vicini* – met in the

18 Simon Keay, 'Decline or Continuity? The Coastal Economy of the Conventus Tarraconensis from the Fourth Century until the Late Sixth Century', in T. F. Blagg *et al.*, eds., *Papers in Iberian Archeology*, BAR International Series 193 (1984): pt ii, 552–77, on p. 566; Philip Banks, 'The Roman Inheritance and Topographical Transitions in Early Medieval Barcelona', in *ibid.*, pp. 600–34, on pp. 600, 610.

19 Joan Francesc Clariana and Marta Prevosti, 'Sobre la pervivencia de hábitats rurales romanos en la Alta Edad Media en el Maresme', *II CAME*, III, 429–36.

20 Antonio Gómez Becerra, *El Maraute (Motril): Un asentamiento medieval en la costa de Granada* (Motril, Ayuntamiento, 1992), pp. 183–7.

Map 2 *Roman settlement on the coast of Granada*

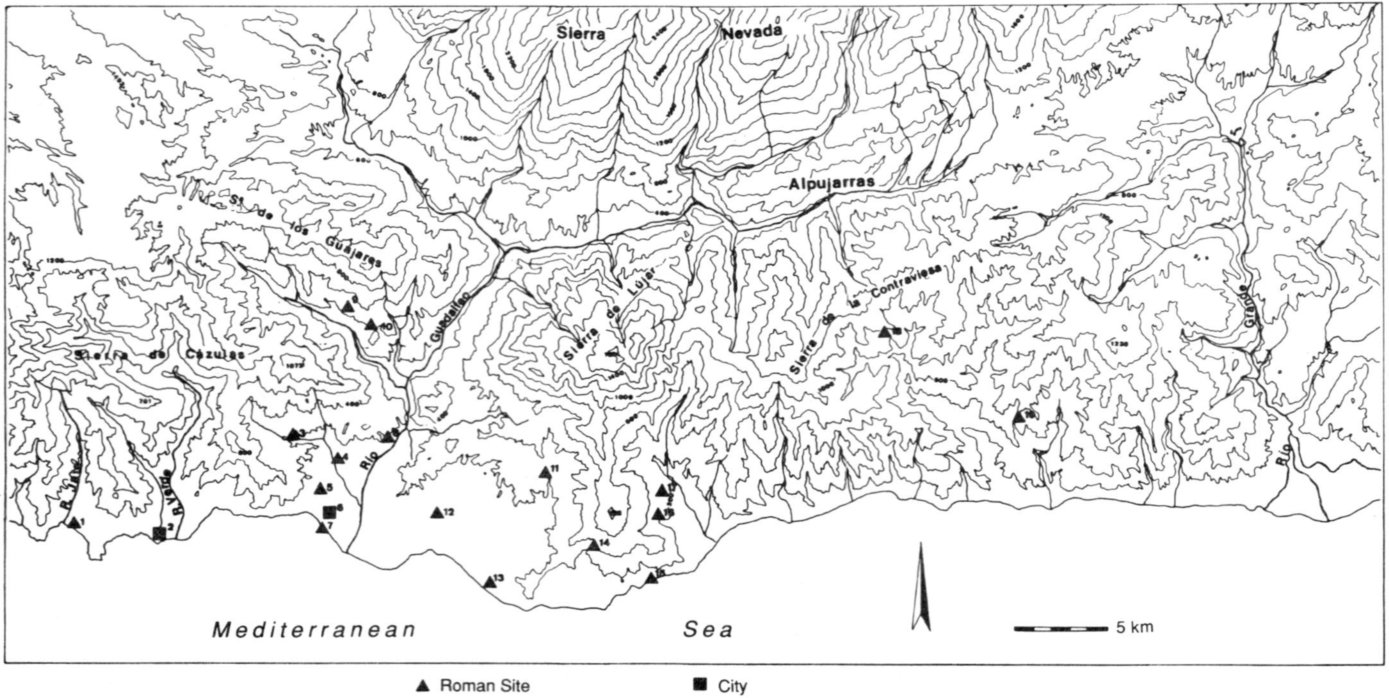

1.- Necrópolis de la Herradura. 2.- Almuñécar. 3.- Yacimiento de la Loma de Ceres (Molvízar). 4.- Yacimiento del Barranco Fortuna (Salobreña). 5.- Yacimiento de Los Barreros (Salobreña). 6.- Salobreña. 7.- Peñón de Salobreña. 8.- Yacimiento del Cerro del Vínculo (Salobreña). 9.- Yacimiento de Los Cortijuelos (Los Guájares). 10.- El Minchar (Los Guájares). 11.- La Herrería (Motril). 12.- Motril (?). 13.- El Maraute (Motril). 14.- Yacimiento del Cortijo de La Reala (Motril). 15.- La Rijana (Gualchos-Castell de Ferro). 16.-Yacimiento de Los Chortales (Gualchos-Castell de Ferro). 17.-Yacimiento de Gualchos. 18.- Yacimiento de la Haza de los Olivos (Bordomarela, Torvizcón). 19.- La Ermita del Palomar (Albuñol).

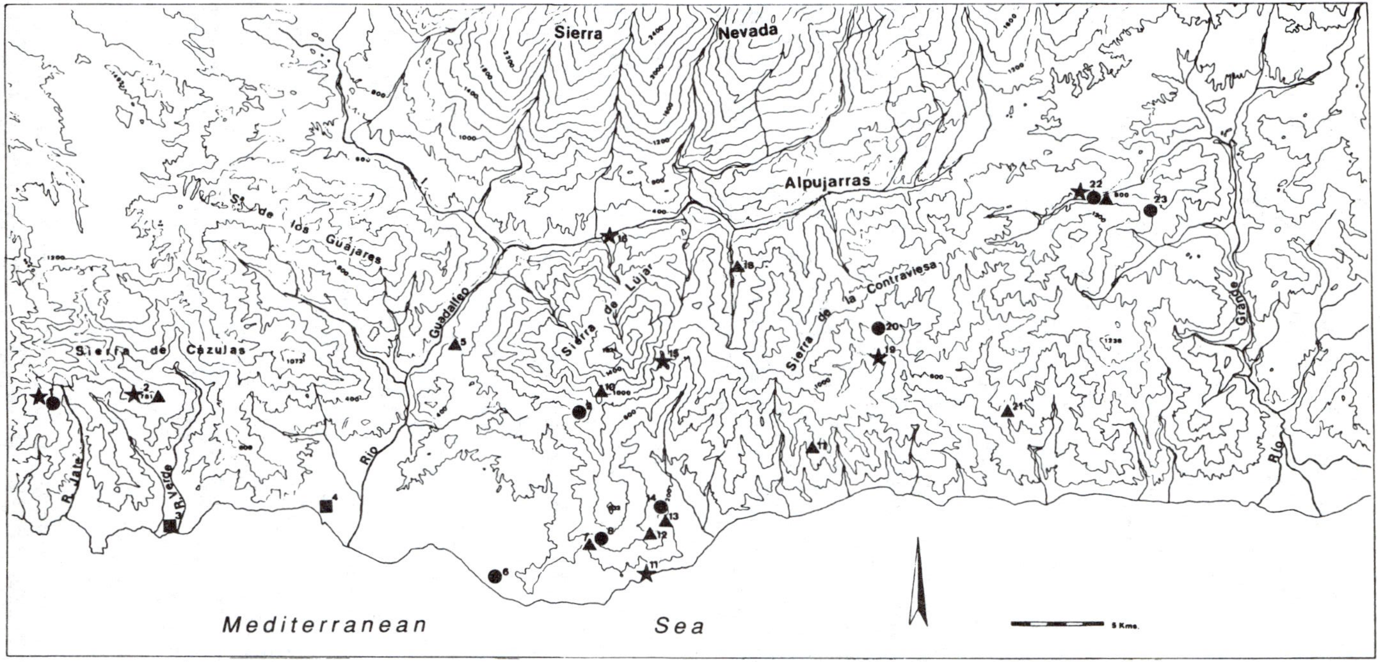

★ Castle (hisn) ■ City ▲ Early Medieval Site ● Caliphal Site

1.- Peñón de Los Castillejos - *Jate* (Almuñécar). 2.- Pico Moscaril (Almuñécar). 3.- Almuñécar. 4.- Salobreña. 5.- Los Castillejos (Vélez de Benaudalla). 6.- El Maraute (Motril). 7.- Yacimiento del Cortijo del Cura (Motril). 8.- Yacimiento del Cortijo de La Reala (Motril). 9.- Yacimiento de La Haza de los Almendros de Lagos (Vélez de Benaudalla). 10.- Picos del Castillejo (Vélez de Benaudalla). 11.- La Rijana (Gualchos-Castell de Ferro). 12.- Pico Águila (Gualchos-Castell de Ferro). 13.- Yacimiento del Cortijo de Los Pastores (Gualchos-Castell de Ferro). 14.- Yacimiento (?) de Gualchos. 15.- Castillo de Olías (Órgiva). 16.- Castillejo de Órgiva. 17.- Los Castillejos (Polopos). 18.- Yacimiento del Peñón de Pedro Vélez (Órgiva). 19.- Castillejo de la Rambla del Valenciano (Sorvilán). 20.- Yacimiento de Bordomarela (Torvizcón). 21.- La Ermita del Palomar (Albuñol). 22.- Castillo y asentamiento de Juliana (Murtas). 23.- Yacimiento de Pino (Murtas).

Map 3 *Medieval Muslim settlement on the coast of Granada*

Gothic equivalent of an Anglo-Saxon folkmoot, the *conventus publicus vicinorum.*[21] Isidore of Seville provides a similar settlement hierarchy of *civitates, castra, villae,* and *viculi* but when such terms appear in documents it is never clear whether they reflect a real hierarchy or simply an appropriation of Isidorean terminology.[22] In documents of the fifth to the seventh centuries, villa no longer meant aristocratic domain and may have just been a village – any village – smaller than a *vicus.* The peasant hamlet and old villa had become indistinguishable, few Visigothic *villae* having more than 100 inhabitants. From the sixth century, churches became focal points of old villas – the future villages of the high Middle Ages. The typical *castrum* or *castellum,* defending strategic points on the Roman road network, may also have been a population nucleus, whether a fortified village or a military edifice with an associated population.[23]

The standard map of Visigothic settlement centred on the meseta in the present province of Segovia was a fabrication of the 1930s and 1940s.[24] Were it true, one would be hard-pressed to account for the population of the Gothic capital of Toledo. Some Visigothic settlement continued the hilltop style so characteristic of late Roman Hispania. Al-Ḥimyarī describes the castle of Bobastro, in the high mountains near Córdoba, as 'a Visigoth district capital, comprising a great number of monasteries, churches and vaulted buildings. It has under its dependence a great number of villages and important fortresses.'[25]

In place of a Visigothic kingdom presumed to have preserved a Goth/Roman cultural dichotomy till its very end, archaeologists have developed a new view which downplays the supposed cultural distinctions (based on the material record, in any case) between Hispano-Romans and Goths and presumes a culturally more heterogeneous picture. The north-west of the peninsula – Galicia,

21 Claudio Sánchez-Albornoz, 'La repoblación del reino asturleonés. Proceso, dinámica y proyecciones', *Cuadernos de Historia de España,* 53–54 (1971), 236–459, on pp. 339–40.

22 José Angel García de Cortázar, 'La repoblación del Valle del Duero en el siglo IX: Del yermo estratégico a la organización social del espacio', in *La reconquista y repoblación de los reinos hispánicos: Estado de la cuestión de los últimos cuarenta años* (Zaragoza, Diputación General de Aragón, 1991), pp. 15–40, on p. 23.

23 Luis A. García Moreno, *Historia de la España visigoda* (Barcelona, Cátedra, 1989), pp. 205–9; Simon Keay, *Roman Spain* (Berkeley, University of California Press, 1988), p. 216.

24 Olmo, 'Historiografía', p. 159.

25 E. Lévi-Provençal, *La Péninsule ibérique au moyen-age d'après le Kitāb ar-Rawḍ al-Mi'ṭār fī Khabar al-Akṭār d'ibn 'Abd al-Mun'im al-Ḥimyarī* (Leiden, E. J. Brill, 1938), pp. 46–7 (trans.). The Visigoths are here called *'Ajam.*

Cantabria and the Basque Country – was much less Romanised than the other regions. In the central zone, formerly viewed as a stronghold of Visigothic settlement and culture (based in large part on erroneous assessment of so-called 'Visigothic necropoli') it can be demonstrated that Romans and Goths shared a common culture from the sixth century on, continuous with that of the Late Empire. Although fine ceramics were produced in revitalised cities like Toledo and Recópolis, built by Leovigild in 578, common unglazed wares modelled by hand or on turntables were associated from the fourth century on with household workshops in hilltop settlements. The third zone comprises the Mediterranean coast, part of the Ebro Valley and the south-eastern corner of the peninsula, those areas under Byzantine rule or in close contact with it, like northern Almería. Although trade with North Africa and the East wound down in Catalonia in the mid-sixth century, it continued in Valencia because of its propinquity to the Byzantine enclave. A fitting symbol is the site at Pla de Nadal, a new villa built in the Visigothic period. In general, however, there was a reduction throughout this zone in the number of *villae* and a consequent consolidation of *fundi*.[26]

Miquel Barceló supposes that Visigothic cities were in an advanced state of decay when the Arabs arrived, yet observes that both Christian and Arab sources describe the conquest of a series of cities and *castra*. Thus the nuclei of Visigothic administration must have survived there.[27] But late Roman urban landscapes, with their distinctive public buildings, were replaced by medieval-looking cityscapes dominated by monasteries and churches. Visigothic cities are associated archaeologically with poor construction and frequent re-use of old materials scavenged from Roman ruins; only churches and bishops' residences were built anew. If one compares the archaeological strata in the city of Valencia for example, that of the High Empire looks red, the Low Empire orange, and the Visigothic, greyish-black (representing rubble and burned buildings). Geo-archaeological analysis of fluvial sediments in the Valencian huerta shows an irrigation regime in imperial times, followed by a rather

26 Following the excellent survey of Lauro Olmo Enciso, 'El reino visigodo de Toledo y los territorios bizantinos. Datos sobre la heterogenidad de la Península Ibérica', in *Coloquio Hispano-Italiano de Arqueología Medieval* (Granada, Patronato de la Alhambra, 1992), pp. 185–98.

27 Miquel Barceló, 'La primerenca organització fiscal d'Al-Andalus segons la "Crònica del 754" (95/713–138/755)', *Faventia*, 1 (1979), 231–61, on p. 247. Arce, 'Transformación de Hispania', p. 247, dissents strongly from my view (following Leopoldo Torres Balbás) that many Roman cities were buried under subsoil when the Arabs arrived (Glick, *Islamic and Christian Spain*, p. 30).

long lapse (fifth and sixth centuries) consistent with the erosion of the canal system.[28]

The revival of the import trade in olive oil and red slip pottery registered between Visigothic towns and Vandal North Africa was not an index of urban prosperity but a symptom of 'the economic dislocation of towns from their surrounding territory'. As coastal towns lost their function as markets for their hinterlands, villas became increasingly independent as economic units.[29]

It is also Barceló's view, however, that Visigothic society was picked off so easily by the Arabs because it had previously suffered a severe agricultural crisis which in turn provoked a demographic contraction and depletion of the labour force between 640 and 680, leading in turn to a diminution in both the tax base and in the rents of the pre-feudal aristocracy. The crisis was touched off by a number of natural catastrophes, including summer droughts and a series of locust plagues extending over a seventy-year period beginning in 583.[30]

28 Simon Keay, *Roman Spain* (Berkeley, University of California Press, 1988), p. 211; Gutiérrez Lloret, 'Poblamiento tardorromano', p. 328 (referring to Elche). Pilar Carmona González, 'Interpretación paleohidrológica y geoarqueológica del substrato romano y musulmán de la ciudad de Valencia', *Saitabi*, 40 (1990), 163–76, on p. 173.

29 Keay, *Roman Spain*, p. 214.

30 Miquel Barceló, 'Les plagues de llagost a la Carpetània', *Estudis d'Història Agraria*, 1 (1978), 67–84.

2

The countryside of Al-Andalus: a new model of settlement

Castles and villages in Al-Andalus

The major achievement of the archaeologists has been to block out the social vertebration of much of rural Al-Andalus and to have described a system of fortifications and their dependent villages, these latter in large part cultivating irrigated fields. The significance of these results is obvious: Al-Andalus was a tributary state; the tribute was paid by peasants. With no understanding of how the latter were organised, both in terms of production and social and political organisation, our understanding of the Islamic polity must be severely one-sided, limited practically to the State itself.

The research was stimulated by the *incastellamento* debate and was carried out in Sharq al-Andalus (today's Valencian community and part of Murcia) by Pierre Guichard (historian) and André Bazzana (archaeologist) and in Almería by Pierre Cressier (archaeologist) and associates, with important essays of interpretation and theorisation contributed by the archaeologist Rafael Azuar Ruiz and the historian/archaeologists Manuel Acién Almansa and Miquel Barceló. With the exception of Guichard, who is known for an earlier volume on the tribal nature of Andalusi rural society, none of these authors is cited in recent English-language writing on Islamic Spain; and yet they are the architects of a major revolution in historical thinking on the subject.

Toubert himself, in prefacing a volume on rural castles in Al-Andalus, provides a historiographical context for such a study and sets out an agenda for future studies of *incastellamento.* First, the kind of archaeology (generally referred to as 'extensive archaeology') is really an archaeology of settlement. From the tenth century on, these fortified habitats played a role that exceeded a simple military one: they constituted 'the structures which contained habitat and organised landholding'.[1] Note that this general definition is applicable to both Christian *and* Islamic worlds.

1 Pierre Toubert, 'Préface' to André Bazzana, Patrice Cressier and Pierre Guichard, *Les châteaux ruraux d'Al-Andalus* (Madrid, Casa de Velázquez, 1988), pp. 9–13, on p. 10.

Toubert continues with a prospective programme for *incastellamento* studies: a more finely-tuned periodisation, a more refined typology of castles and systems of castles, and, perhaps most importantly, a focus on the relationship between fortified habitats and the villages that gravitated around them, their articulation with secondary defence structures (such as watch-towers, or *atalayas*), and associated phenomena of irrigation and water use.[2] Looking at Europe generally, Toubert sees the dating of *incastellamento* as problematic. Were Latin *castra* and Muslim *ḥuṣūn* both called into being in the ninth century to assure 'public' control of an early phase of rural growth and settlement, or do they date to the tenth and eleventh centuries and represent a response to the disaggregation of central power?[3]

As we will have occasion to note in some detail, the *ḥuṣūn* of Al-Andalus were not in any way 'feudal' installations nor were they part of a society which was in any way 'feudal'. Nevertheless, in a general framework of transition from late antiquity, Toubert presumes a broad chronological and *functional* similarity in process in the Islamic and Christian worlds.

The flight of population from lowlands to the mountains, a general phenomenon in the western Mediterranean world of late Roman times, led in Spain to a reoccupation of hilltop or hillside strong points – some of them the sites of ancient Iberian *oppida*, and many such sites have been identified from the sixth and seventh centuries.[4] Such hilltop sites were also significant in the first two Islamic centuries in Spain, which is why many of the *ḥuṣūn* established in this period had Latin names: that is, the social role constituted by the castle and its territory crystallised in the very early Middle Ages, before linguistic Arabisation had had a chance to occur. Those with Arabic place-names (such as names compounded on *qala*, Alcalá, 'the fortress', for example) are newer and lower down. Thus in Cocentaina, the Pic Negre site, established in the fifth or sixth century, was abandoned for the tenth-century Muslim-built *ḥiṣn*, now called El Castell.[5] In the Alpujarras, while

2 *Ibid.*, pp. 10–11.

3 *Ibid.*, p. 12.

4 The castle/refuge (*castella tutiora*) was known in late Roman times, as when the inhabitants of Gallaecia took refuge from the Suevi in such; Javier Arce, 'La transformación de Hispania en época tardarromana: Paisaje urbano, paisaje rural', in *De la Antigüedad al Medioevo, siglos IV–VIII. III Congreso de Estudios Medievales* (Madrid, Fundación Sánchez-Albornoz, 1993), pp. 225–49.

5 Josep Torró, *Poblament i espai rural: Transformacions històriques* (Valencia, Diputació Provincial, 1990), pp. 44–7; Pierre Guichard, 'El problema de la existencia de

most *ḥuṣūn* are not documented in Arab histories until the tenth century, two (Juliana and Escariantes) are mentioned by al-Udhrī as having played a role in an uprising against the Emir Hisham I around 788. Earlier *ḥuṣūn* were of the *ḥiṣn*/refuge type with an empty enclosure (see below) and no permanent occupation. Then, the *fitna*, or civil strife, of the late ninth century produced a further movement of *incastellamento*. These later *ḥuṣūn*, which formed a political and strategic system of castles, were more complex, capable of sustaining permanent garrisons and sometimes a small peasant population.[6]

The basic unit of rural settlement as described by the archaeologists is a castle/village (*ḥiṣn/qarya*) complex. The castle itself, whatever else it might have, usually had a central redoubt called *salūqiya* in Arabic,[7] *celoquia* in Romance, where the *qā'id* resided. Beyond this redoubt was a walled-in space called *baqqar* in Arabic, *albacar* in Romance, frequently quite large and usually provided with a cistern, where the rural population took refuge in times of war or unrest. For example, the very large *ḥiṣn* of Gartx (Bolulla, Alicante) had a tripartite scheme consisting of a hilltop castle (the *celoquia*), refuge (the *albacar*), and a village within a third enclosed space; that of Chivert (as described in Latin in its Carta Puebla) had a *ḥiṣn* in two parts, a mosque, a cistern with a collection system, an *albacar* with houses in it, and a suburb (*arrabal*); that of Alcalá or Benisilí (Vall de Gallinera), a *celoquia* and two *albacar*/s.[8]

Dependent on the castle for the use of the *albacar* as a refuge were a number of villages (Arabic, *qurā*; singular *qarya; alquería* in Romance). There is ample evidence from right after the Christian conquest demonstrating that Muslim peasants were habituated to

---

estructuras de tipo "feudal" en la sociedad de Al-Andalus', in P. Bonnassie *et al.*, *Estructuras feudales y feudalismo en el mundo mediterráneo* (Barcelona, Crítica, 1984), pp. 117–45, on p. 129.

6 See Bazzana's typology of *ḥuṣūn*: 'Typologie . . . : Les habitats fortifiés du Sharq al-Andalus', in *Habitats fortifiés et organisation de l'espace en Méditerranée médiévale* (Lyon, Maison de l'Orient, 1983), pp. 19–27, on p. 27. See also Valérie Dallière-Benelhadj, 'Le "Château" en Al-Andalus: Un problème de terminologie', *ibid.*, pp. 63–7; Bazzana, Cressier and Guichard, *Châteaux ruraux*, p. 130; Vicente Salvatierra Cuenca, *Cien años de arqueología medieval. Perspectivas desde la periferia: Jaén* (Granada, Universidad, 1990), p. 92.

7 *Salūqiya* originally referred to the cabin of a ship's pilot; see R. Dozy, *Supplement aux dictionnaires arabes*, 2nd ed., 2 vols. (Leiden, 1927), I, 676.

8 (Gartx) Bazzana, Cressier and Guichard, *Châteaux ruraux*, p. 79; (Chivert), Guichard, *Musulmans de Valence*, I, 211; Rafael Azuar Ruiz, *Denia islámica: Arqueología y poblamiento* (Alicante, Diputación Provincial, 1989), p. 77. See Latin references to *albacar*/s in Philippe Sénac, 'Poblamiento, habitats rurales y sociedad en la Marca Superior de Al-Andalus', *Aragón en la Edad Media*, 9 (1991), 389–401, on p. 396 nn. 23 (*illo albacar de illo castello*), 24.

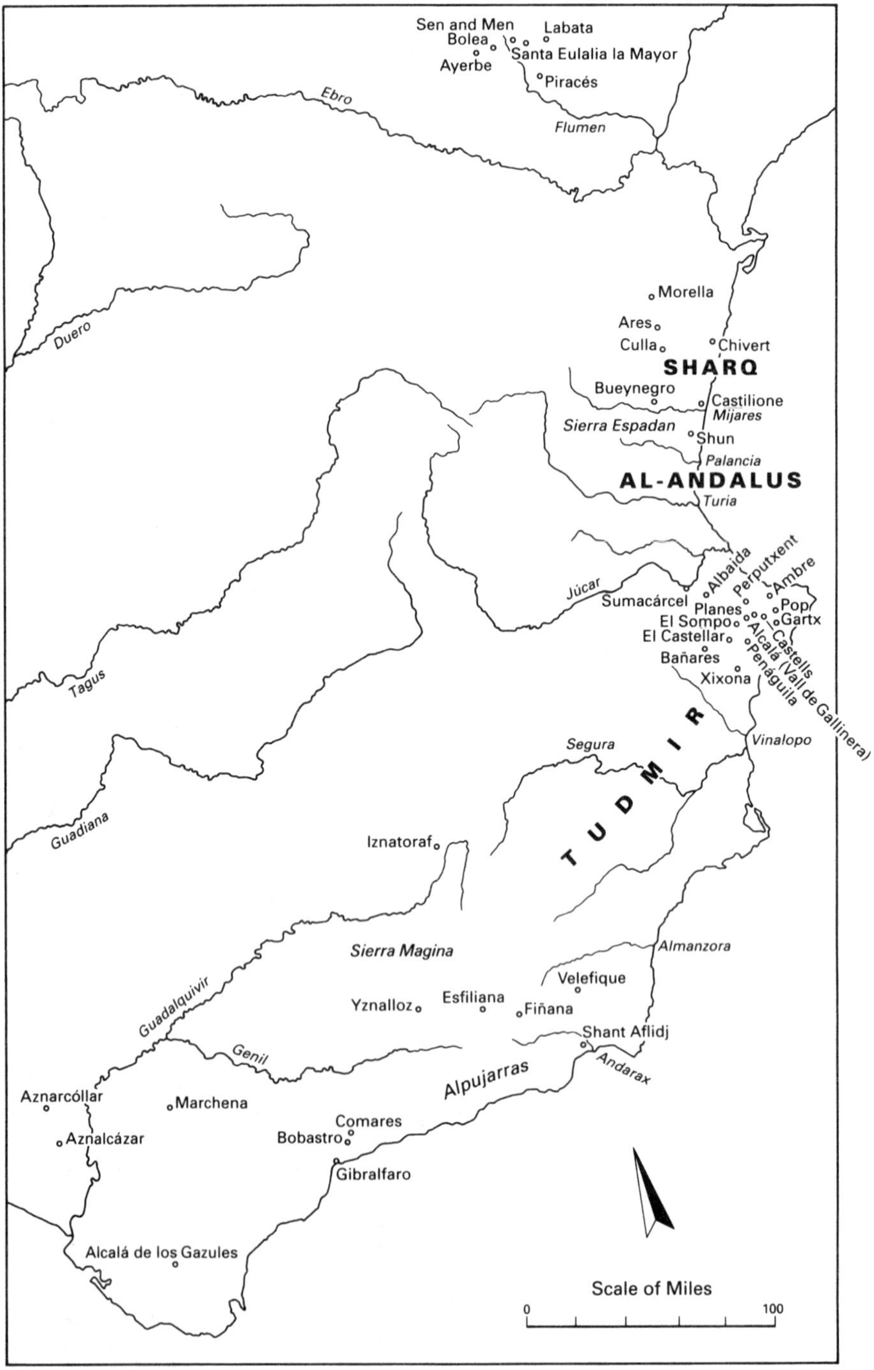

Map 4 *Ḥuṣūn (castles) in Al-Andalus*

using the *ḥuṣūn* as refuges, even when Christians controlled the latter. Thus Bernat Desclot reports that during the Mudéjar revolt of 1276–77, the Muslims 'emptied out their villages in the plains and went up with the beasts and clothing to the feet of the castle walls'.[9]

The *qā'id* (the terms of whose appointment are not clear and no doubt changed over time) was the Government's representative, whose main roles were to collect taxes and direct the defence of the locale. The villagers were freemen and had no feudal obligations to the *qā'id*, although they had a civic obligation called *shukhra*, which consisted in transporting water and firewood to the castle. Besides the villages, which were, at least in principle, settled by tribal segments or clans and farmed collectively, there were private estates called *raḥal*/s in some places, particularly in the environs of cities.

When one then looks at *systems of castles*, slightly different patterns can be detected. In the Huesca region of the Upper March, where the Christian threat was a constant of daily life, the twelve places identified as *ḥuṣūn* in Arabic sources constituted a defensive system centred around Huesca, the principal *madīna*. These include a defensive perimeter of five northern-lying fortresses of Ayerbe, Bolea, Sen and Men (occupying two peaks on either side of the River Flumen), Santa Eulalia la Mayor, and Labata.[10] Like other *ḥuṣūn*, these differed in architectural detail, but all had a refuge space near the summit and, in the lower parts of the castle, many rooms. The relationship between these defensive installations and agriculture has not yet been clarified. Although each *ḥiṣn* controlled an area of some 100 sq km, it is not clear whether they were nodal points of agrarian colonisation or whether there was a peasantry previously in place.[11]

In the north of Valencia, the present-day province of Castelló, there appears to have been no functional distinction between frontier *ḥuṣūn* and those of the interior. Indeed, the *ḥiṣn* system there seems to have disregarded the frontier almost entirely: in the early thirteenth century, the 80 km of frontier with Aragón were defended

9 Cited by Josep Torró, 'El problema del hábitat fortificado en el sur del reino de Valencia después de la segunda revuelta mudéjar (1276–1304)', *Anales de la Universidad de Alicante. Historia Medieval*, 7 (1988–89), 53–81, on p. 55.

10 See the map of *ḥuṣūn* and *alquerías* in the Huesca region in Carlos Laliena and Philippe Sénac, *Musulmans et Chrétiens dans le Haut Moyen Age: Aux origines de la conquête aragonaise* (Toulouse, Minerve, 1991), p. 69.

11 Philippe Sénac and Carlos Esco, 'Le peuplement musulman dans le district de Huesca (VIIIe–XIIIe siècles)', in *La Marche Supérieure d'Al-Andalus et l'Occident Chrétien* (Madrid, Casa de Velázquez, 1991), pp. 51–65. The authors do not mention *alquerías* at all.

by but three *ḥuṣūn* (Culla, Ares and Morella), while the marginally shorter frontier with Catalonia merited four. The density of castles was the greatest in the south, where the bulk of the population lived, that is, on the plain of Castelló, in the valleys of the Mijares and Palancia rivers, and so forth. Where the population was densest, so were the castles. Bazzana and Guichard point out that the Christians were unable to perceive any difference between frontier and interior castles in the Valencian region, while for Ibn al-Abbār, who knew the region well, the former commanded frontier regions – *thugūr* – while the latter's jurisdiction was the *ᶜamal*, the standard administrative district. The territories therefore must have been organised differently, but there is no evidence as to how.[12] What can be surmised from this distribution of castles is, first, that defensive considerations were not limited to frontier areas: by the late twelfth century at the latest, everyone felt threatened. This would explain why the huge *ḥiṣn* of Bueynegro does not appear to reflect the defensive requirements of its modest rural population.[13] Second, whatever the specific strategic or military function the castle might have had, the main role as a node of local social organisation was unimpaired. Butzer notes that in Castelló *qarya/*s came to be fortified in the twelfth century in response to the Christian threat. But his conclusion that 'by 1200 there probably was little difference between the typical *ḥiṣn* and the typical *qarya*' is hyperbolic.[14]

The castle/village complex

In Al-Andalus the *alquería* was the basic fiscal unit on which state finances were based.[15] A sum of 7–10 *alquerías* for each *ḥiṣn* seems to have been the norm. Al-Udhrī gives 148 *ḥuṣūn* and 1,079 *alquerías* for the *kura* of Córdoba, while al-Zuhrī, for the Sierra de Segura (Albacete) gives figures of 33 *ḥuṣūn* and 300 *alquerías.*[16] For Córdoba, therefore, there were 7.3 *alquerías* for each *ḥiṣn,* in Albacete, 9.1. The size of *alquerías,* as measured by Torró in Alcoi, ranged

12 André Bazzana and Pierre Guichard, 'La frontière du Sharq al-Andalus', in *La Marche Supérieure d'Al-Andalus et l'Occident Chrétien,* pp. 77–88.

13 *Ibid.*, p. 87. On the pervasive sense of threat, see comment by Pedro Chalmeta in the same volume, p. 88.

14 Karl W. Butzer, 'Castles on the Valencian Border March', *Al-ᶜUsur al-Wusta: Bulletin of Middle East Medievalists,* 4:2 (October 1992), 17–19, 39, on p. 19.

15 See Miquel Barceló, 'La primerenca organització fiscal d'Al-Andalus segons la "Crònica del 754" (95/713–138/755)', *Faventia,* 1 (1979), 231–61.

16 Manuel Acién Almansa, 'Sobre la función de los ḥuṣūn en el sur de Al-Andalus. La fortificación en el Califato', in *Coloquio Hispano-Italiano de Arqueologia Medieval* (Granada, Patronato de la Alhambra, 1992), pp. 263–74, on p. 226; Bazzana, Cressier and Guichard, *Châteaux ruraux,* p. 61.

between 72 and 90 ha, quite close to the mean of 83.72 calculated by Poveda for Mallorca.[17] But, of course, these figures are based on the successor Christian *alquerías* which, unlike their Andalusi antecedents, had measured metes and bounds. Cressier (among others) has stressed the difficulty in reconstructing the territory of an *alquería*. According to this author, the only case in which one can be reasonably sure that the Christian *alquería* corresponds to its Islamic predecessor is that of a middling irrigation system (mesosystem) where there is a clear relationship between the entire *alquería* and the irrigation canal or canals. In a dry-farming region, it is impossible to determine the extent of the Islamic village.[18] But beyond such practical considerations, there is a conceptual one: the Christian concept of a village, with private (or even collective) parcels, measured and recorded, responding to the property rights of an individual or corporate body, did not exist in Andalusi *qurā*, where the undivided, customarily defined district was held to be the collective property of a tribal segment. *Alquerías* were tribal or clan space, sometimes not even nucleated but consisting of a number of small settlements.[19] Most *alquerías* were small, with a handful of houses in each, and agricultural land worked at least in part collectively. Burns styles these *alquerías* 'amorphous'; I would say they were unbounded, but not necessarily amorphous, because that would imply that the inhabitants did not have a concept of the geographical bounds of the village. However, I think the *Repartimiento* evidence discussed in Chapter 6 shows that the Muslim residents of *alquerías* knew where their limits (in a customary, even jurisdictional, sense) were, although such conclusions should not be applied anachronistically to the thirteenth century. In any case, Burns's evidence shows both sides of the coin: that *alquerías* had no fixed boundaries, but their

17 Josep Torró, *Alcoi. La formació d'un espai feudal (de 1245 a 1305)* (Valencia, Diputació Provincial, 1992), p. 45; Angel Sánchez Poveda, 'Introducción al estudio de la toponimia árabe-musulmana de Mayurqa según la documentación de los Archivos de la Ciutat de Mallorca (1232–1278)', *Awraq*, 3 (1980), 75–102, on p. 96. The aggregate figure from Mallorca, Menorca and Ibiza is 85.5 ha for *alquerías*, 49.6 ha for *raḥal*/s ('Toponimia árabo-berber', p. 601). Pierre Guichard and André Bazzana, 'La sociedad musulmana valenciana en vísperas de la conquista cristiana', *Nuestra Historia* (Valencia, Mas Ivars, 1980), II, 263–80, on p. 271, notes that *alquerías* in irrigated districts were small (0.5 to 2.5 sq. km), while in the mountains they were quite large (Benilloba, 9.26 sq km, Benasuau, 9.5, and so forth).

18 Patrice Cressier, 'Agua, fortificaciones y poblamiento: El aporte de la arqueología a los estudios sobre el sureste peninsular', *Aragón en la Edad Media*, 9 (1991), 403–27, on p. 411.

19 Miquel Barceló, 'Vísperas de feudales. La sociedad de Sharq al-Andalus justo antes de la conquista catalana', in Felipe Maíllo Salgado, ed., *España, Al-Andalus, Sefarad: Síntesis y nuevas perspectivas* (Salamanca, Universidad, 1988), pp. 99–112, on p. 106.

residents knew where the jurisdiction was. Therefore in Xiu (Vall d'Albaida) the owners of certain fields testified that they were 'not in the district of the said village, nor did the said village in Saracen times have fixed boundaries'. The King concluded that 'any village [*alquería*] of the kingdom of Valencia has no fixed boundaries [*non habet terminos certos*] except only those boundaries which the Saracens of the said village were accustomed to work, returning from them the same day from their plot to that village'. We will observe later that by the sixteenth century or even earlier many *alquerías* had acquired marked boundaries while still in Muslim hands. A lawsuit in Bizar (now Policar, Granada) at the end of the fifteenth century produced testimony to the effect that the *alquería* had distinct and well-known lands and limits in Muslim times. Such limits were not necessarily marked off physically, however. In Olula del Río (Almería), the bounds of the village had not been established until 1564, according to the *apeo* of 1572.[20]

It is clear from the evidence marshalled by Guichard that *alquerías* were in no way seigniorial; they were not the domain of anyone, they had no proprietors other than the tribal collectivities after whom they were frequently named, nor did they evolve from the Roman *vicus* or villa, at least not in a social sense.[21]

If the individual *alquería* was, from a modern perspective, weakly structured, the 'castral territory' (*ḥiṣn/qarya* complex) was sharply structured.[22] Particularly in mountainous regions such as Alicante or the Alpujarras of Granada where the typicality and ubiquity of

20 Robert I. Burns, *Muslims, Christians, and Jews in the Crusader Kingdom of Valencia* (Cambridge, Cambridge University Press, 1984), pp. 215–23. Josep Torró, 'Sobre ordenament feudal del territori i trasbalsaments del poblament mudèjar. La *Montanea Valencie* (1286–1291)', *Afers*, 7 (1988–89), 95–124, on p. 104; Manuel Espinar Moreno, 'Bizar: Una alquería musulmana y el paso al domino cristiano (siglos XII–XVI)', in *Andalucía entre Oriente y Occidente (1236–1492)* (Córdoba, Diputación Provincial, 1988), pp. 707–18, on p. 714. José Domingo Lentisco Puche, *La repoblación de Olula del Río (Almería) en el siglo XVI* (Almería, Instituto de Estudios Almerienses, 1991), p. 45.

21 Pierre Guichard, 'Le problème des structures agraires en Al-Andalus avant la conquête chrétienne', in *Andalucia entre Oriente y Occidente (1236–1492)* (Córdoba, Diputación Provincial, 1988), pp. 161–70. On pp. 164–5, he criticises the feudalising interpretation of *qarya*/s by previous historians, including Julio González, Jaime Oliver Asín, Manuel Sanchis Guarner, and Robert I. Burns. For González, the *alquería* has an 'eminent proprietor'; Oliver Asín stressed structural continuity in the dominial structure of landholding from Roman Spain through Islamic times and continuing to the present as 'the farm-hamlet of an important landlord'.

22 It seems clear from Idrisī's description of castles, according to Roberto Marín-Guzmán, that 'a fortified place was named *ḥiṣn* if it did not have a densely concentrated population' of *alquerías* which it protected ('The Revolt of 'Umar Ibn Ḥafṣūn: A Challenge to the Structure of the State (880–928)', unpublished doctoral dissertation, University of Texas at Austin, 1994, p. 260).

the phenomenon is clearest, the system included a castle or refuge and seven to ten irrigated *alquerías*. The minimal *ḥiṣn/qarya* complex would supply the basic daily needs of the peasantry (as enumerated by Cressier): defence and protection in the castle, religion (in one or more mosques), and agriculture (an irrigation system watering one or more *alquerías*). Interstitial spaces used for sheep herding, some wheat farming, and exploitation of forest or other wild resources, complemented the core complex.[23] In some cases, the *ḥiṣn*, or perhaps the central village associated with it, was the seat of an administrative district or *juz*.[24]

Recent research has tended to demonstrate the universality of the *ḥiṣn/qarya* complex throughout Al-Andalus, with the exception of Mallorca. Thus in Portugal, archaeological reconnaissance of Alcaria Ruiva, in the Andalusi *kūra* of Beja, on the Alvacar (= *albacar*) river (indicative of an old *ḥiṣn*) and irrigated by a spring, displays all the typical features of the *ḥiṣn/qarya* complex.[25] In the Algarve as well, the *ḥiṣn/qarya* complex seems to have been the characteristic form of settlement.[26] In the mountains east of Jáen, thirteenth-century documents reveal the same kind of settlement pattern.[27]

The religious function of the *ḥiṣn/qarya* complex can be surmised from a number of texts and some complementary archaeological evidence. A rural mosque at Velefique studied by Angelé and Cressier reveals some of the dynamics of these complexes. Velefique was the central community among fourteen *alquerías*, whose inhabitants performed Friday prayer in its mosque. During the civil unrest (*fitna*) of the eleventh century, Friday prayer was performed in the mosque of the *ḥiṣn* itself. Similarly, al-Udhrī mentions a Friday mosque in the *ḥiṣn* of Piracés in the Upper March. According to Ibn al-Khaṭīb, there were fifty *alquerías* with Friday mosques in the Nasrid kingdom.[28]

23 Cressier, 'Agua, fortificaciones y poblamiento', pp. 413–14.

24 Marín-Guzmán, 'Revolt', p. 265.

25 Mercè Argemi *et al.*, 'Alcaria Ruiva: Un assentament rural a l'Alentejo', *Arquelogia Medieval* (Portugal), in press.

26 Christophe Pica, 'L'Evolution des localités de l'Algarve du XIème au XIIIème siècles', *Cahiers d'Histoire*, 37 (1992), 3–21, esp. pp. 9–16.

27 Milagros Jiménez Sánchez and Tomás Quesada Quesada, 'En los confines de la conquista castellana: Toponimia y poblamiento de los montes granadino-gienneneses en el siglo XIII según la documentación cristiana', *Revista del Centro de Estudios Históricos de Granada y su Reino*, 2nd epoch, 6 (1992), 51–80, on pp. 66–9. See also José A. Rodríguez Lozano, 'Ḥiṣn Yiliana > Esfiliana', *Miscelánea de Estudios Arabes y Hebráicos*, 41–2 (1991–92), 337–45, esp. p. 343, citing the *alquerías* that formed this particular complex.

28 Sabine Angelé and Patrice Cressier, 'Velefique (Almería): Un ejemplo de mezquita rural en Al-Andalus', in Cressier, *Estudios de arqueología medieval en Almería* (Almería, Instituto de Estudios Almerienses, 1992), pp. 241–63. Sénac, 'Poblamiento, habitats

Little is known about what was actually grown in such rural *alquerías*. Butzer's evidence suggests that the 'Arab Green Revolution' never reached many rural places, whose cultivators continued to grow the same range of cultivars as their Roman antecedents, with no trace of the roster of Indian and Persian crops, such as rice, cotton, and oranges, introduced in the peri-urban huertas of the great cities.[29]

Zozaya has found that the general distribution of settlements in Al-Andalus obeyed a logic of distance, with large settlements spaced every 30 km, with *alquerías* or inns every 15 km, the distance that can be walked in a day.[30] This notion is well supported in the literature of 'central place theory', which supposes that the hierarchy of settlements obeys a logic determined by the economic division of labour between small, middling and large centres, the latter providing services to the former in return for agricultural produce.

Especially in the vicinity of towns, there were also private estates, called *raḥal*/s, typically the property of important personages. Of 150 agricultural places in the huerta of Valencia mentioned in the *Repartiment*, two-thirds were *alquerías*, one-third *raḥal*/s, also cultivated, but of much less extent. They have personal names or that of a functionary (e.g., Raḥal al-Qāḍi, the judge's estate), while *alquerías* tend to have the names of kinship groups. In Mallorca, 62 per cent of agricultural units were *alquerías*, 35 per cent *raḥal*/s; in Menorca the figures are 75 per cent and 17 per cent respectively, while in Ibiza, *alquerías* accounted for 81 per cent of the territory, *raḥal*/s for 19 per cent. The mean size of *raḥal*/s was 49.02 ha (versus 83.72 for *alquerías*). The *Repartimiento* of Murcia shows very large *raḥal*/s, such as an unirrigated one of 59 ha on the edge of the huerta probably belonging to the *rais* of Orihuela. Some Murcian *raḥal*/s were sited on irrigation canals but on land too high to water; others were irrigated with lifting devices. In any case, *raḥal*/s always tended to be geographically marginal because of the

rurales', p. 397. Vincent Lagardère, *Campagnes et paysans d'Al-Andalus (VIIIe–XVe s.)* (Paris, Maisonneuve et Larose, 1993), pp. 88–9. Marín-Guzmán, 'Revolt', p. 296 n. 2.

29 Karl W. Butzer *et al.*, 'Irrigation Agrosystems', pp. 502–3. The persistence of a pattern of Roman cultivars supports our notion of Paleoandalusí culture in rural Al-Andalus (see Ch. 3). Butzer, in his conclusions, makes no allowance for regional variation.

30 Juan Zozaya, 'Notas sobre las comunicaciones en Al-Andalus omeya', *II CAME*, I, 219–28, on pp. 226–7.

impossibility of breaking into the tightly organised, sharply structured, kinship-based irrigation communities of the huerta.[31]

Arabists have attacked virtually every aspect of Guichard's construction of the social significance of the *ḥiṣn.*[32] According to Rubiera and Epalza, the *ḥuṣūn* were part of a defensive system oriented towards the sea; it failed when the enemy approached from the interior. Thus it was obsolete by the end of the twelfth century, when the Almohads built huge castle–fortresses. They assume that *ḥuṣūn* were instruments of central or at least urban power, that they had tax collectors (the *ᶜamīl,* fiscal officer of an administrative district, *ᶜamal*), serving under a *walī* living in a city.[33] The *ḥiṣn*'s role was as a tax-collecting centre, to control the peasantry, not to protect them, and so the *albacar* was not a refuge but rather a place to store the tribute, which was typically paid in kind, in the form of livestock. (Although *baqar* is indeed Arabic for cattle, we need not inevitably expect to find cattle there, any more than we would expect to find a ship's pilot in a *celoquia.*) In this sense, the *albacar* is analogous to the state grain depot, the *hury*, reflected in place-names like Alforín, Alorín, Alfolí.[34] The *albacar* was simply a holding pen for cattle; it could not be defensive because of the weakness of the walls and the low height (90 cm) of some of them. Moreover, an archaeological survey by Màrius Bevià of the *ḥiṣn* of Alicante produced cattle droppings![35] To establish an economic function for

31 Pierre Guichard, *Les Musulmans de Valence et la Reconquête (XIe–XIIIe siècles)*, 2 vols. (Damascus, Institut Français de Damas, 1990–91), II, 375–86; Angel Poveda Sánchez, 'Sobre los distritos, las explotaciones y la toponimia clánica de Yabisa (Eivissa)', *Sharq Al-Andalus,* 1 (1984), 109–15, on p. 112; Angel Poveda Sánchez, 'Toponimia árabe-musulmana de Mallorca', p. 96, and 'Toponima àrabo-berber i espai social a les Illes Orientals d'Al-Andalus', unpublished doctoral dissertation, Universitat Autònoma de Barcelona, Bellaterra, 1987, pp. 599, 601.

32 In a review of Pierre Guichard, *Structures sociales 'orientales' et 'occidentales' dans l'Espagne musulmane* (Paris, Mouton, 1977), the French edition of *Al-Andalus: Estructura antropológica de una sociedad islámica en Occidente* (Barcelona, Barral, 1976), Peter Linehan noted that Guichard's attack on traditional historiography was so broad that 'there is no knowing where the counterattack will reassemble. But reassemble it surely will'; *Social History,* 3 (1978), 377–9, on p. 379. The counterattack, as here described, in fact reassembled under the banner of Arabism. Traditional medievalists were oblivious of Guichard's relevance to their interests and, in large part, still are.

33 In this regard, cf. Angel Poveda Sánchez's finding of an inverse spatial relationship between tribal settlement in Mallorca and political power: tribal settlements clustered in administrative districts farthest away from the seat of government at Madīnat Mayūrqa [= Palma]; 'Toponimia àrabo-berber', p. 595.

34 María Jesús Rubiera and Míkel de Epalza, *Xàtiva musulmana (segles VIII–XIII)* (Xàtiva, Ajuntament, 1987), p. 26.

35 Míkel de Epalza, 'Funciones ganaderas de los albacares, en las fortalezas musulmanas', *Sharq al-Andalus,* 1 (1984), 47–54; Màrius Bevià, 'L'Albacar musulmà del castell d'Alacant', *Sharq al-Andalus,* 1 (1984), pp. 131–40.

the *albacar* is important to Epalza's view of a centrally organised tributary system completely controlled by and from urban centres. Guichard's response is that Christian documentation of the thirteenth century makes clear their defensive function, which continued to even later times, for example, the famous congregation of Moriscos in Pop in 1609. And certainly why would people want to bring their cattle to inaccessible places like the *ḥuṣūn* of Pop or Penáguila?[36]

According to the Arabist critique, Guichard also misrepresents the nature of the *qarya*. These had family names (in Beni-), not because they were the collective property of an agnatic group, but because they were owned by an absentee proprietor. The villagers were his sharecroppers. The place-names are plural (Beni- and not Ibn) because the properties stayed in the same family over many generations.[37]

The *raḥal*, moreover, is not a private estate, but rather a simple country hut used by itinerant shepherds. Guichard's distinction between the free proprietors of the *alquería* and the seigniorial owners of *raḥal*/s did not exist, according to Rubiera. The debate over the meaning of *raḥal* is instructive because it illustrates the kinds of difficulties of interpretation involved in the reconstruction of the Andalusi countryside from Christian documents. Thus in medieval Valencian documents, *rafal* could either mean *raḥal* or mill (*raḥa al-*) or even suburb (*raval*, from Arabic *rabaḍ*). The matter is further complicated by confusion with *real*, an unwalled garden, from Arabic *riyāḍ*.[38]

Here Guichard rejoined that in the *Repartiment* of Mallorca, the Arabic text shows the *raḥal* linked with personal names; that in the *Repartiment* of Valencia *raḥal*/s in the Valencian huerta were agricultural estates, with both irrigated and dry-farmed components;

36 Guichard, *Musulmans de Valence*, I, 214–16; Bazzana, Cressier and Guichard, *Châteaux ruraux*, pp. 29f. On Pop, see Pierre Guichard, 'El castillo y el valle de Pop durante la edad media: Contribución al estudio de los señoríos valencianos', *Anales de la Universidad de Alicante. Historia Medieval*, 2 (1983), 19–31.

37 Rubiera and Epalza, *Xàtiva musulmana*, pp. 27–31.

38 María Jesús Rubiera, 'Rafals y reales; Ravals y arrabales; Reals y reales', *Sharq al-Andalus*, 1 (1984), 117–22. Guichard, *Musulmanes de Valence*, II, 379–87. Robert I. Burns, *Foundations of Crusader Valencia* (*Diplomatarium*, II) (Princeton, Princeton University Press, 1991), p. xiii, would authorise the whole range of possible meanings: *rafal* means 'sheepfold, a shepherd's rest, or a stockraising rural property, developing later the meanings of a rural shack, country place or even farm'. Following Rubiera, he translates *rafal* as 'livestock pen' (*ibid.*, p. 283) or 'cattle place' (p. 301). A hapless translator from Catalan represents *rafal* as 'outhouse'! (*Estudi General*, 5–6, p. 560).

thus, in this limited sense, they may well have also served as enclosures for cattle; and that in *Repartimiento* of Murcia, *raḥal*/s are on the edge of the huerta, are large and unirrigated, used not for grazing but for dry-farming of cereals.[39] In my view the cattle-pen solution is inherently implausible: it is clear, particularly from the *Repartiment* of Mallorca, that the Muslim rural settlement there was very nearly bimodal: if not an *alquería*, then a *raḥal*.

The discussion of the *raḥal* has to be placed in the context of Al-Andalus as a whole. In the kingdom of Granada, there were similar privately-owned, typically large estates called *majshar*/s, while in the Upper March *almunia*/s were similarly rural domains belonging to a sole proprietor. In addition to *almunia*/s, Sénac has located seventeen *raḥal*/s or *rafales* in the province of Huesca.[40] Moreover, the same division was characteristic of Muslim settlement in Sicily, where the settlement units were either *ḍayʿa*/s (as in Spanish *aldea*) or *raḥal*/s; the two terms are virtually synonymous. Hence: Raḥal Indulcini (the farm of the Andalusis), Raḥalstephani, Raalgermani, Raḥalnicola, Raḥalkerames, and so forth. It is interesting that *raḥal* here is compounded with personal or family names, as had been the style in Arabic. There was also a Norman lord there named Guillaume de Raḥaltawyl (the Tawīl being a prominent Arab tribe). This family of magnates would hardly take their name from a cattle pen![41]

Finally, to emphasise the point that Guichard has misconstrued the entire nature of rural society, Epalza and Rubiera argue that the *sukhra* or *sofra* – peasant work services owed to a castle – was not a general imposition associated with all *ḥuṣūn*, but only appears in

39 Pierre Guichard, 'El impacto de la Reconquista en la sociedad musulmana', in *Historia del Pueblo Valenciano* (Valencia, Levante, 1988), pp. 221–40, on p. 228; 'A propos des raḥals de l'Espagne orientale', *Miscelanea Medieval Murciana*, 15 (1989), 11–24. Robert Pocklington presumes the *raḥal* to have been dry-farmed, by definition, and translates the term 'cortijo de secano': *Estudios toponímicos en torno a los orígenes de Murcia* (Murcia, Academia Alfonso X El Sabio, 1990), pp. 225–6.

40 Amador Díaz García and Manuel Barrios Aguilera, *De toponimia granadina* (Granada, Universidad, 1991), pp. 227–8; Laliena and Sénac, *Musulmans et Chrétiens*, p. 68. Sénac, 'Poblamiento, habitats rurales', p. 400. According to Lagardère, *Compagnes et paysans*, p. 57, there is no (functional) distinction between *raḥal*/s and *munya*/s.

41 Henri Bresc, 'Féodalité coloniale en terre d'Islam: La Sicile (1070–1240)', in *Structures féodales et féodalisme dans l'Occident méditerranéen (Xe–XIIe siècles): Bilan et perspectives de recherches* (Rome, Ecole Française de Rome, 1980), pp. 631–47, on pp. 635 n. 9, 640 nn. 29 and 31, 642. The term is used in an identical sense in Malta: Miquel Barceló, Joan Pinyol and Angel Sánchez Poveda, 'Eren ramaders els rafals de Mayurqa? Un exercici de simulació històrica', in *Les Illes Orientals d'Al-Andalus* (Palma de Mallorca, Institut d'Estudis Baleàrics, 1987), pp. 115–22, on p. 120 n. 5.

mountain zones.[42] This is an odd kind of argument, inasmuch as a plurality of *ḥuṣūn* were in the mountains. Burns's data indicates that the core obligation of the *sofra* was to provision castles with water and wood. This obligation was generalised throughout Al-Andalus and survived until the fifteenth century. Thus the residents of the *alquerías* in the zone of the *ḥiṣn* of Comares (Málaga) were obliged, according to the *Repartimiento* of 1487, to supply work, water and building materials for the fortress, even though the King paid the master masons.[43]

Two writers, Rafael Azuar Ruiz and Manuel Acién Almansa, have criticised the general model, Azuar Ruiz from the perspective of his own archaeological studies of *ḥuṣūn* in Alicante, Acién Almansa from a more theoretical and synthetic perspective. Azuar Ruiz agrees that the *ḥiṣn* is a socially-defined entity: when Arab travellers say they stopped in the *ḥiṣn* of such-and-such a place, they mean something more than a castle. Indeed it is a broad entity and can be composed of a castle and a variable number of *alquerías* and *raḥal*\s.[44] It is the political meaning of the *ḥiṣn* that most concerns him. A *ḥiṣn* has to have a *qā'id* and vice versa. Guichard has made the *qā'id* into a secondary figure and he exaggerates the administrative autonomy of rural communities, or *aljamas.* The *qā'id* is an important official so long as the central power, whose delegate he is, exists; in such times, the *alquerías* depend on the *ḥuṣūn,* not the other way around. The *qā'id* loses that attribute when the power disappears, leaving only the *aljama* and the tribal governance structure of *shaykh*\s (called *vells* in late medieval Catalan documents).[45]

In a recent book expounding a much fuller set of findings covering the Taifa state of Denia, Azuar Ruiz argues for late, defence-related *incastellamento* in his study area. The earliest castles (including that of Denia, El Castellar of Alcoi, and Pic Negre and El Sompo of Cocentaina) date to the end of the tenth century, with only Pic Negre displaying remains from an earlier epoch. Even *ḥuṣūn* founded on Iberian or Roman sites, like El Castellar and Denia must be reckoned as new foundations, as Muslims took advantage of local

42 Míkel de Epalza and María Jesús Rubiera, 'La *sofra* (*sujra*) en el Sharq al-Andalus (s. XIII)', *Sharq al-Andalus,* 3 (1986), 33–8.

43 Robert I. Burns, *Medieval Colonialism: Postcrusade Exploitation of Islamic Valencia* (Princeton, Princeton University Press, 1975), pp. 162–73; Francisco Bejarano-Robles and Joaquín Vallvé, eds., *Repartimiento de Comares 1487–1496* (Barcelona, Universidad, 1974), p. 6.

44 Rafael Azuar Ruiz, 'Una interpretación del "ḥiṣn" musulmán en el ámbito rural', *Revista de Estudios Alicantinos,* 37 (1982), 33–41, on p. 38.

45 *Ibid.*, p. 39.

'habitat traditions'.[46] A second phase responds to a deliberate policy of military construction under the Almoravids, dating to the second half of the eleventh century. Examples are the *ḥuṣūn* of Ambre (Pego) and Planes, characterised by a ramp leading to an elbow-bend entrance, defended by an ante-wall. Azuar Ruiz doubts these are *albacar* castles: they are not refuges, he argues, but rather fortified settlements in depopulated areas whose function was to control roads or passes (e.g., the *ḥiṣn* of Jijona).[47]

The third phase, in which the greater part of the castles were constructed, corresponds to the Almohad building programme of the end of the twelfth and beginning of the thirteenth centuries. Those castles with the best preserved Islamic features (Bañeres, Biar, Jijona) all have a *tapia*-wall tower of three floors, surrounded by a walled enclosure. The *ḥuṣūn* linked up with a system of towers which exercised a 'coercive function' over the administrative territory of the castles and protected its communication system.[48]

The castles of Azuar Ruiz's study area therefore are easily recognised, having polygonal floor plans, toothed walls, and ramp entrances with an elbow-bend. They were built with normalised methods (e.g., standard *tapia* – rammed earth – frames and polygonal towers analogous to the Christian *donjón/s*) and they represent a technological advance with respect to earlier, hilltop sites. In Almohad castles, defence was supplied by the thickness and height of the walls, not by the natural defences of remote hilltop sites. These technological improvements, he adds, have nothing whatever to do with local tribal organisation. Therefore, *incastellamento* was *late* in southern Alicante and was related to the coercive objectives of the Taifa state and Berber dynasties, defined by the buoyant economies of the *taifas* and by the fiscal requirements of those governmental entities.[49]

Therefore Azuar Ruiz, while admitting the plurifunctionality of *ḥuṣūn* (including the protection of groups of *alquerías*), assigns much greater significance to their military and defensive roles than do Bazzana and Guichard.

The latter's reply is that Valencian *ḥuṣūn* generally are an expression of segmentary sociopolitical organisation, fragmented into relatively autonomous rural communities. These date to earlier

46 Rafael Azuar Ruiz, *Denia islámica: Arqueología y poblamiento* (Alicante, Diputación Provincial, 1989), pp. 338–40.

47 *Ibid.*, pp. 340–2.

48 *Ibid.*, pp. 343–7.

49 *Ibid.*, pp. 348–9.

times and, for hilltop sites such as El Castellar (Alcoi), Sumacárcel and Uxó, there is ample evidence to support an early foundation. They go on to restate the rationale of *incastellamento* studies generally: the traditional view of European castles is that they were built because of fear of foreign invaders. The new view stresses their association (in Christian Europe) with feudalism, in Islamic Spain, with rural social and agrarian organisation.[50] In segmentary societies, moreover, fortifications were built by tribal groups to defend against other tribes.

Acién Almansa's critique is of particular interest from the perspective of the Roman to Islamic transition and makes clear that Guichard's theorisation of rural social organisation provides a useful point of departure for any further theorisation. Moreover, Acién Almansa's interest is in widening the compass of *ḥuṣūn* studies to include not only Berbers, but the mass of indigenous inhabitants (*muwalladūn*) as well. Acién Almansa's view is that, in Andalusia, part of the indigenous population took advantage of the conquest to flee to the high country (to escape the fiscal exactions to which they had been accustomed) and reoccupy abandoned hilltop sites: the ancient Iberian *oppida*. The heirs of the old aristocracy also fled to the mountains where they built more complex fortresses, called *Ummahāt al-ḥuṣūn*, with larger populations, a stronger fortification (*alcazaba*), and suburbs. These are controlled by *ṣāḥib*/s, some of whom are *muwallad*/s of Gothic ancestry. These castle/settlements had a double function: they were centres for taxing villages and refuges for banditry practised against Islamised centres. The Valencia-style refuge *ḥuṣūn* also existed here, the reaction of tribal society against state control and against the threat of the *ṣāḥib*/s of the Ummahāt. Ibn Ḥayyān described in some detail ᶜAbd al-Raḥmān's campaign to wipe out independent foci of rebellion: 'the entire *kūra* of Rayya was left without any fortified mountain'; or how, in Sidonia, he established people on the plains and destroyed the *ḥuṣūn*.[51]

The double process of fortification, particularly in Andalusia, permits the explanation of the different typology and function of the larger, defence-oriented *Ummahāt al-ḥuṣūn*, on the one hand, and the *ḥuṣūn*-refuges, on the other, which are adaptations to the social development of the indigenous population, though not limited to it.[52] Acién Almansa's construction of the 'country of *ḥuṣūn*',

50 Bazzana, Cressier and Guichard, *Châteaux ruraux*, pp. 32–8.

51 Acién Almansa, 'Función de los *ḥuṣūn*', p. 265.

52 Manuel Acién Almansa, 'Poblamiento y fortificación en el sur de Al-Andalus. La formación de un país de *ḥuṣūn*', *Actas, III CAME*, I, 135–50, on p. 145.

therefore, has the advantage of linking up a putative Berber style of segmentary social organisation with the mass of indigenous peoples. In this sense, the creation of this distinctive landscape is not so much owing to the wholesale transplantation of Berber social organisation from North Africa where, at least in the case of Morocco, there is no evidence for the kind of refuge fortresses as they appear in Al-Andalus,[53] but rather to a distinctive adaptation of the immigrant tribal groups to a movement of indigenous settlement set off by the invasion and conquest of the country in 711.

We end up with Al-Andalus described as a 'country of *ḥuṣūn*'. Is this accurate? With the exception of Mallorca, undoubtedly so. But it is interesting to note that the *ḥiṣn/qarya* system, while seemingly Magribī in that Berbers were heavily implicated in its organisation, does not correspond with any system as yet revealed by extensive archaeology of Morocco, where rural fortresses are notably lacking. The conclusion that one might draw is that the *ḥiṣn* system of Al-Andalus drew upon and extended a pre-existent system of hilltop habitats which was characteristic of the Eastern Spanish landscape at the time of the eighth-century conquest. This is the pattern suggested by the fact that the preponderance of *ḥuṣūn* have Latin names, while *alquerías* typically have Arab or tribal names. The *ḥiṣn/qarya* system was in place by the ninth century. Before then the situation requires clarification: we might well surmise an extended period of the structural reorganisation of the countryside.[54]

These critiques are part of a process of working out the fine texture, both functionally and chronologically, of Guichard/Bazzana's universalising hypothesis. Functional distinctions emerge the better we know the historical circumstances of the *ḥiṣn* in question, nor can we doubt that certain castles were refitted in times of extended military activity.

Berbers in Al-Andalus

There is no doubt that the ethnic distribution of settlement is an important factor in understanding the society and culture of Islamic Spain.[55] It is especially important in conceptualising the transition from a Roman to Muslim organisation of the countryside, because different ethnic groups had distinctive agricultural practices. Berbers

53 Acién Almansa, 'Función de los ḥuṣūn', p. 269 n. 11, citing Cressier on Morocco.

54 Bazzana, Cressier and Guichard, *Châteaux ruraux*, pp. 296–7.

55 See my discussion of ethnic relations in Al-Andalus in *Islamic and Christian Spain in the Early Middle Ages: Comparative Perspectives on Social and Cultural Formation* (Princeton, Princeton University Press, 1979), pp. 178–85.

were pre-eminently herdsmen and tree growers, but also practised irrigation both in mountain and plain environments. The propinquity of the Magrib and the presumption that more Berbers than Arabs migrated to Spain obliges us to weigh the Berber contribution. Guichard and Barceló have insisted on the importance of Berber settlement in Sharq al-Andalus and Mallorca respectively. But they, and anyone else supporting a Berberising hypothesis, have been attacked by Arabists who claim, first, that the evidence for Berber settlement is not as clear as those authors would like and, second, that the weight of Berber culture in Andalusi society was very small. Those who so argue assume that Andalusi Berbers displayed a high degree of Arabisation.

Eduardo Manzano argues that most of the first wave (eighth century) of Berber settlers were Butr tribes from Tripolitania, who were not Arabised nor even fully Islamised (some may well have been Berber-speaking Christians), who settled, in particular, on the northern frontier. (The tribal groups in question were the Maṭgara, Madyūna, Hawwāra, and Miknāsa.)[56] Peter Scales, reassessing the weight of Berber settlement in the first century of Andalusi history, accepts Manzano's argument that early Berber settlers were mainly from Butr tribes.[57] He takes Guichard to task for assuming the significance of tribal structures in the early centuries when – says Scales – *ᶜaṣabiyya* (Ibn Khaldūn's 'group feeling' or solidarity) was personal or political, not tribal or racial.[58] Guichard has equated the *qawm* or tribal segment with Ibn Khaldūn's *ahl* – the group in general.[59] But then Scales does not take the next step and examine the logic of Guichard's reconstruction of the Berber landscape on the basis of place-names. Thus his critique is peculiar, intimating that the early Berber settlers did not function as Berbers because, according to Scales's interpretation of Ibn Khaldun, if Berber tribal *ᶜasabiyya* does not exist, then neither does the Berber.[60]

One of the most distinctive, and controversial, attributes of the *ḥiṣn/qarya* system is the toponymic record it left behind. Frequently

56 Eduardo Manzano, 'Beréberes de Al-Andalus: Los factores en una evolución histórica', *Al-Qanṭara*, 11 (1990), 397–428, on pp. 418–27; and *La frontera de Al-Andalus en época de los omeyas* (Madrid, CSIC, 1991), pp. 234–7. Manzano has attacked the conventional use of the term as something more than a 'lineage' (*Frontera*, p. 44). However, this cavil does not detract from Barceló's use of such data to demonstrate the relationship of settlement groups in different parts of the peninsula.

57 Peter C. Scales, *The Fall of the Caliphate of Cordoba: Berbers and Andalusis in Conflict* (Leiden, E. J. Brill, 1994), p. 146.

58 *Ibid.*, p. 158.

59 *Ibid.*, p. 155.

60 *Ibid.*, p. 143.

Table 1 *Some alquerías with Beni- names*

Name	*Location*	*Derivation*	*Comment*
Benicanena	Gandía	Banū Kināna	Qaysī Arbas
Benicasim	Castelló	Banū Qāsim	Kutāma Berbers
Benicayz	Ibiza	Banū Qays	Arabs
Benifadale	Ibiza	Banū Faḍāla	Arab tribe from Cyrenaica
Benifilell	Ibiza	Banū Hilāl	Arabs
Benigafull	Sagunto	Banū Gafūl	Berber clan
Benigazló	Vall d'Uxó	Banū Gazlūn	Nafza clan
Beniouara	Ibiza	Banū Hawwāra	Berber confederation
Benimazoch	Ibiza	Banū Marzūq	Berber tribe
Benipater	Ibiza	Banū Butr	Berber confederation
Benirroym	Ibiza	Banū Ru'ayn	Yemeni Arabs
Benisanó	Valencia	Banū Zannūn	Hawwāra clan
Benisomada	Ibiza	Banū Sumata	Nafza Berber segment

(and most characteristically in Sharq al-Andalus) the *alquerías* grouped around a particular castle will have a preponderance of names beginning with the element Beni- (Arabic, *Banū,* 'sons of'). These names are typically conjoined with Arab proper names, although the distribution of these place-names does not correspond to areas of heavy Arab settlement. The Beni- names are of unimportant places: villages or quarters of towns or villages, and are typically associated with rural, frequently mountainous places, indicative of settlement by a tribal segment or clan (Arabic, *qawm*). Thus the castral district of Penáguila (Alicante) included the surviving villages of Benilloba, Benifallim, and Benasuau, but also four *despoblados,* two of them with Beni- names as well: Benigama and Beniaf. The castle of Pop (Alicante) likewise had seven *alquerías,* three with Beni- names (Benigela, Benalcabar, Benilacruci, in a document of 1239).[61] The incidence of such names in zones with other place-names identifiable as Berber suggests to Guichard the Berber provenance of these places.[62]

Identifying the personal or tribal name linked with the Beni- particle, however, is a task that runs from the obvious to the highly speculative. It is virtually impossible to establish antiquity or stability of an individual place-name. Guichard's method is inferential: link a place-name with textual evidence of that tribe's presence in

61 Bazzana, Cressier and Guichard, *Châteaux ruraux,* pp. 71, 85.
62 Guichard, *Al-Andalus,* pp. 423–40.

Al-Andalus, or in Valencia, at some specified date. Guichard gives some literary references (late) to *alquerías* with Beni- names, e.g., Qarya Banī Riyāḥ near Ronda, where Ibn Baṭṭūṭa had resided in the house of the *shaykh* of that name.[63]

Mallorca too appears to have been settled mainly by Berbers after its delayed conquest in the early tenth century: of fifteen tribal groups identified by Poveda Sánchez in a study of the place-names found in the *Repartiment* of Mallorca, eleven are Berber (Ghumāra, Maṭgara, Nafza, Hawwāra, Malīla, Banū Zannūn (Hawwāra segment), and Zanāta); four are Arab (Banū Kināna, Quraysh, Bāhila, and Banū Hilāl). Of all tribal place-names, 85 per cent of those in Mallorca are Berber, 100 per cent in Menorca, and 58 per cent in Ibiza. With respect to *alquerías* with Beni- names, Poveda Sánchez established that they cluster around the mean dimension, a highly significant finding which suggests their typicality as a unit of exploitation.[64] Although high concentrations of Beni- names are characteristic of Mallorca and the Valencian region, they appear elsewhere too. Of fifteen toponyms in a small study area in the Sierra de Filabres studied by Cressier and García Latorre, only four do not contain the form Beni-. These were mountain villages, typically divided into two or more small quarters (*barrios*), which was a reflection of the clan structure underlying these settlements.[65]

Like so many missing links, place-names complementary to Guichard's thesis keep turning up. The importance of this toponymic sleuthing cannot be underestimated: the same tribal names appear repeatedly throughout the peninsula, reflecting patterns of migration and segmentation. Thus the Berber Beni Ajjar appear in Lleida, Castelló, Valencia, Alicante, Mallorca, Menorca, and Seville.[66] Likewise, a clear pattern of duplication or replication of clan names in Valencia and Mallorca, indicative of the migration between those two

63 *Ibid.*, pp. 413, 437f.

64 Poveda Sánchez, 'Toponimia árabe-musulmana', pp. 79–80, 95; Angel Poveda Sánchez, 'Toponimia àrabo-berber', p. 594. Beni- place-names account for 20 per cent of place-names in Mallorca, 30 per cent in Menorca, and 41 per cent in Ibiza (*ibid.*, p. 596).

65 Juan García Latorre, 'Arqueología medieval e historia moderna en el reino de Granada. El caso de la Sierra de Filabres', *Chronica Nova*, 20 (1992), 177–207, on p. 182. On 'polynuclear' villages, see Marie-Christine Delaigue, 'Mutations de l'espace villageois en Andalusie orientale. Effects immédiats et lointains de la Reconquête', *Mélanges de la Casa de Velázquez*, 26 (1990), 131–62, on pp. 147–9.

66 See maps in Barceló, 'La cuestión septentrional. La arqueología de los asentamientos andalusíes más antiguos', *Aragón en la Edad Media*, 9 (1991), 341–53, on pp. 351 (Beni Labīd, Gelida), 352 (Iraten/Artana, Beni Ajjar), and 353 (Beni ʿArus). See also Sergi Selma Castell, 'Toponimia tribal i clànica d'origen berber al nord de Sharq al-Andalus (Recull i noves prepostes)', *Estudis Castellonencs*, 5 (1992–93), 459–66, on p. 465.

places of clans. Thus, from Beni Ryagel, Vinarragell near Borriana, Beniraçkel in Mallorca; Benimarva, in Castellón, Benimarvan (Tarragona) and Abenmarwam (Mallorca), all from Benī Marwān, a Ghumāra Berber segment; Artana (Castelló) and Yartan (Mallorcan administrative district), from Berber Iraten.[67] This kind of evidence, in my opinion, establishes both the incidence and consistency of Berber settlement as well as the reality of segmentation and its association with internal migration, and thus constitutes the most conclusive demonstration thus far presented of the Berberisation of the countryside.

In a recent study of southern Catalonia, Miquel Barceló finds a number of place-names derived from tribes in the Alt Penedès: Mediona, from the Madyūna Berbers, a widely dispersed Berber group, which settled at many points along the eastern coast of Spain, and Lavit, from the Banū Labīd, Hilālī Arabs.[68] In the region of Tortosa are two imported Berber place-names: Bitem, after the river Biṭām, near Tobna (Tunisia), which al-Bakrī describes as irrigating a region similar to that of Bitem, on the Ebro; and Mianes, from Mayānish, a place near al-Mahdīya, both places characterised by irrigation with norias. There is, in the Sierra de Espadán, a village named Lauret, named after Lurīt, near Tlemsen: both are irrigated by four springs.[69] More such instances will surely emerge.

Critiques of the Berberisation hypothesis have come mainly from Arabists.[70] Carmen Barceló Torres attacks Guichard's methodology of matching Beni- names with real or supposed Berber figures from Valencia or elsewhere. Since there are no systematic studies of Andalusi personal names, Guichard's hypothesis is purely speculative. In hypercritical fashion, she believes that the Beni- names,

67 *Ibid.*

68 Miquel Barceló, 'Assentaments berbers i àrabs a les regions del nord-est d'Al-Andalus: el cas de l'Alt Penedès (Barcelona)', in *La Marche Supérieure d'Al-Andalus et l'Occident Chrétien*, pp. 89–97, especially pp. 90, 92.

69 Miquel Barceló, 'Aigua i assentaments andalusins entre Xerta i Amposta (s. VII–XII)', *II CAME*, 2: 413–20.

70 Curiously, although attacks on Guichard by Arabists date from the 1980s and 1990s, he had anticipated virtually every one of these arguments in his 1975 book (*Al-Andalus*, pp. 420ff.). Recently he has responded to his critics in a series of essays: Pierre Guichard, 'Faut-il en finir avec les berbers de Valencia?', *Al-Qantara*, 11 (1990), 461–73; 'La toponymie tribale berbère valencienne: Réponse à quelques objections philologiques', in *Festgabe für Hans-Rudolph Singer* (Frankfurt, 1991), 125–41; Pierre Guichard, 'Els Berbers de València, una vegada més. Resposta a Carme Barceló', *Afers*, 15 (1993), 225–32. With personal names, he observes, one must be content with probabilities. While each place-name might be criticised, it is the *ensemble* (that is, the pattern) that interests him. He now believes that Beni- names based on Hawwāra, Madyūna, and Nafza Berber settlement are early, Zanāta and Ṣanhāja probably later.

which can only be documented from the thirteenth century, cannot be associated with historical personages of the ninth century whose only link is purely coincidental.[71]

Epalza and Rubiera also attack Guichard's toponymic evidence. His error, they say, is the anachronistic one of using thirteenth-century documentation to establish eighth-century settlement. They even suggest that the Beni- names may be a mere epiphenomenon of the way in which royal officials received evidence at local inquests when preparing the *Repartiment*. On this view Benicasim would just mean 'Qāsim's men' – Qāsim being the head man of the place – and no tribal affiliation can be assumed. They further criticise some of the etymological base of Guichard's reasoning, such as the confusion between Beni and Benna (= Spanish *peña*, mountain or peak); thus Benisid turns out to mean 'mountain of the governor', rather than 'sons of'. Rubiera also doubts the derivation of various Adzaneta/s from the Berber tribal name Zanāta, and so forth.[72]

Azuar Ruiz, for his part, does not doubt either the social or ethnic significance of *alquerías* with Beni- names; but he is insistent that they refer to a late settlement by Berbers in the second half of the twelfth century, when there was a population spurt as a result of Almohad policy of resettling refugees from the Upper March as well as new arrivals from North Africa.[73]

Although Valencia and to a lesser extent Mallorca are still rich in Beni- place-names, the *Repartimientos* of Andalusia reveal a medieval landscape there with many more such names as currently survive: as a result of resettlement there was a wholesale replacement of Arabic *alquería* names with Romance ones. The picture now emerging makes clear the incidence of Beni- names from all periods

71 Carmen Barceló Torres, '¿Galgos o podencos? Sobre la supuesta berberización del País Valenciano en los siglos VIII y IX', *Al-Qantara*, 11 (1990), 429–60. On the association of Beni- names with other place-names indicating Berber tribal settlement (Guichard, *Al-Andalus*, pp. 437f.).

72 Míkel de Epalza and María Jesús Rubiera, 'Estat actual dels estudis de toponimia valenciana d'origen àrab', *Xé Col·loqui General de la Societat d'Onomàstica. Ier d'Onomàstica Valenciana* (Valencia, 1986), pp. 420–6; María Jesús Rubiera, 'Toponimia arábigo-valenciana: Falsos antróponimos beréberes', in *Miscel·lània Sanchis Guarner* (Valencia, Universidad, 1984), I, 317–20. But there is no inherent reason why the 'philological' method is more likely than the historical one, to produce certainty. In the elucidation of Arabic-derived toponyms, there must be some guesswork: the method requires the interpretation of Arabic words and sounds deformed by the Christian scribes who wrote them down, and there is no way to determine the kind of mediation that took place, its quality, or how many such mediations there were.

73 Azuar Ruiz, *Denia islámica*, p. 419. On Tagarino place-names in the vicinity of Benidorm, see María Jesús Rubiera and Míkel de Epalza, *El noms àrabs de Benidorm i la seua comarca* (Alicante, Universitat, 1985).

and not exclusively late, as Epalza argues. Were they all late, they would not be so widely distributed throughout the peninsula. Nor, as Acién Almansa observes, are they limited to a single ethnic group. Thus the Beni Maurell were Muwallad, the Beni Hilāl, Arab, and so forth. Still the ubiquity of Beni- names 'implies a common social ambiance, very possibly the distribution of lineages and irrigated lands suggested by Barceló'.[74] A landscape rich in Beni- names is distinctively North African, today's Moroccan gazetteer displaying 150 settlements so named, that of Algeria, 85.[75]

Archaeological evidence, which thus far is sparse, may bolster one hypothesis or the other. Thus Acién Almansa observes that late Roman pottery forms do not appear in the areas of Berber settlement in the Serrania of Ronda, and goes on to say that the absence of such forms in Valencia supports Guichard's Berberisation hypothesis. But Gutiérrez Lloret's findings from southern Alicante, including the important site of the *Rābiṭa* (*a rābiṭa* or *ribāṭ* was a frontier outpost for ascetic soldiers) of Guardamar (see Map 5), throws into question the intensity of the Berberisation of Valencia, at least the southern region.[76]

My own view is that while it is true (as Carmen Barceló has observed[77]) that a place-name derived from a tribal name does not mean that only persons of that tribe lived there, there is evidence in plenty to substantiate the tribal basis of *alquería* settlements. But that is really begging the issue: indeed very late evidence (1391) shows that in Benirrama (Alicante), for example, nine of nineteen heads of family were still named 'Ibn Raḥma'.[78] Such places, to the extent possible under Christian rule, still attempted to organise their social life according to norms of tribal governance.

It is interesting to note that the polemic of the Beni- names replicates an equally violent debate over Germanic place-names that took place a few generations ago. As Marc Bloch explained:

74 Manuel Acién Almansa, 'Recientes estudios sobre arqueología andalusí en el sur de Al-Andalus', *Aragón en la Edad Media*, 9 (1991), 361.

75 *Morocco*, United States Board on Geographic Names (Washington, DC, 1970), pp. 148–59; *Algeria* (Washington, 1972). I have counted only settlements (= PPL); there are also many tribal areas (= TRB) so listed.

76 Manuel Acién Almansa, 'Cerámica a torno lento en Bezmiliana. Cronología, tipos y difusión', *I CAME*, IV, 243–67, on p. 248; Sonia Gutiérrez Lloret, *Cerámica común paleoandalusí del sur de Alicante (siglos VII–X)* (Alicante, Caja de Ahorros Provincial, 1988), p. 190.

77 Barceló Torres, '¿Galgos o podencos?', p. 447.

78 Guichard, *Musulmans de Valence*, I, 227 n. 23. As late as the fifteenth century, in Andalusia, where the hold of endogamy had long since been broken, there were still *alquerías* where a substantial portion of the population was still descended from the founding *qawm*.

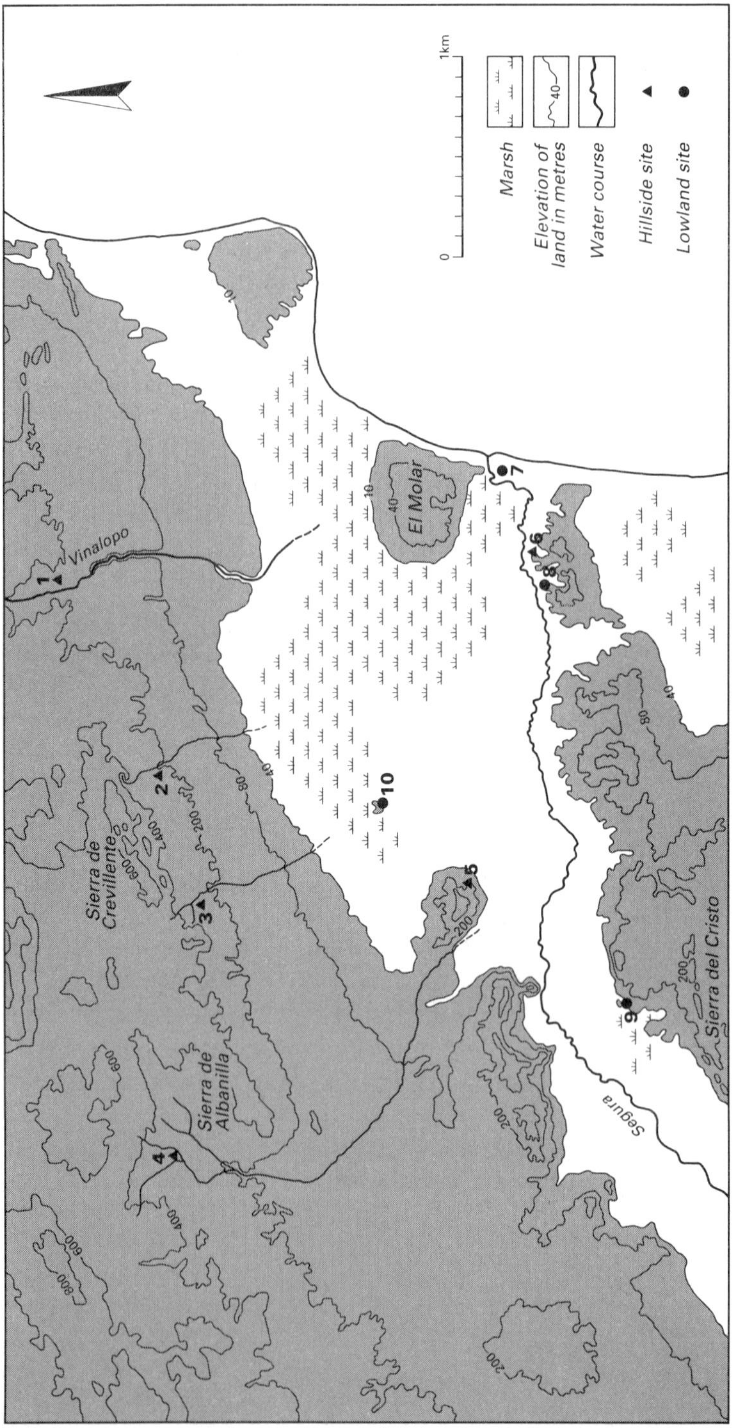

Map 5 *Paleoandalusi settlement in the lower Segura basin, 8th–10th centuries*

Germanic countries show native personal names with various suffixes, of which the oldest are in -ing and in –heim. (The old view that the -ing suffix implied tribes or clans has given way to the view that it merely implies any sort of dependence; the Heuchlingen may be Huchil's men or his relatives, perhaps both.) . . . [place] names of this sort exist almost everywhere.[79]

The view that *-ing* place-names may represent either tribal or non-tribal dependence relations is more or less what Epalza suggests for Beni- names. And of course it may well be that, by the time the Christians recorded these data, the names of villages had long since stabilised and the persons there had no relation to the founding *shaykh* or his tribe, or had perpetuated such relationships fictively. Evidence presented below in Chapter 6 shows that clan organisation persisted late enough to support the tribal reality of Beni- names.

Those who have attacked the Berberisation hypothesis have also attacked Guichard's construction of the social meaning of the *ḥuṣūn/qarya* complex: we have considered Epalza's critique in particular. But Guichard's model of settlement does not *require* Berbers, just segmentary social organisation. Such settlements could as well be Arab or, if Acién Almansa's thesis proves acceptable, Muwallad as well.

79 Bloch, 'The Rise of Dependent Cultivation and Seignorial Institutions', in *Cambridge Economic History of Europe*, 2nd ed. (Cambridge, Cambridge University Press, 1966), I, pp. 235–90, on p. 272.

3

The Paleoandalusi period: a new culture emerges

Historical inferences from the ceramic record

The new archaeology has generated various schemes of periodisation based on the archive of material culture – mainly pottery – whose historical utility requires validation. Indeed, it is a rationale of the present volume to suggest the relevance of archaeological findings to historical synthesis and to specify an appropriate level of interaction between the two fields.

In a pioneering synthesis of Andalusi pottery which sought to establish criteria for classifying forms both geographically and chronologically, Retuerce Velasco and Zozaya coined the term 'Paleoandalusi' to characterise wares of the Umayyad emirate (roughly from the mid-eighth century to the end of the ninth century).[1] Then, Sonia Gutiérrez Lloret, in an important study of plain, unglazed pottery in the south of the present province of Alicante, from the seventh to the tenth centuries, used the same term to characterise common wares of the early Islamic period, so strikingly different from the well-studied glazed wares of the caliphal period.[2] The area studied was a homogeneous, highly Romanised area embracing the Segura and Vinalopó basins. This is an area characterised by the persistence of late Roman forms well into Islamic times. The Visigoths had no ceramics of their own; thus Hispano-Roman forms continued through the Visigothic period. These were 'modelled' pots, either hand-built or created using a turntable, a wheel which turns too slowly to permit the piece to be raised and is viewed as an aid to hand-building. The turntable can be used intermittently, the potter moving the wheel towards herself with her left hand, modelling with the right, or it can be used more like a wheel, spinning the turntable with the left hand and using both hands to model. Hand-built forms predominate through the

1 Manuel Retuerce Velasco and Juan Zozaya, 'Variantes geográficas de la cerámica omeya andalusí: los temas decorativos', in *La cerámica medievale nel Mediterraneo occidentale* (Florence, Edizioni all'Insegna del Giglio, 1986), pp. 69–128, on p. 70.

2 Sonia Gutiérrez Lloret, *Cerámica común paleoandalusí del sur de Alicante (siglos VII–X)* (Alicante, Caja de Ahorros Provincial, 1988).

ninth century, reproducing a characteristic phenomenon of late antiquity. Gutiérrez Lloret does not interpret hand-building as a symptom of cultural lag or of ignorance of the kickwheel: the two techniques coexisted.[3]

In the Islamic sites studied, 13 per cent of the pots were thrown, 87 per cent hand-built, and more than half of the specimens are large boiling pots, called *marmitas.* This is a late Roman form with a flat base, found in this area from the sixth and seventh centuries to the eighth centuries, with the mouth closing up over time. With the addition of an internal glaze, the form lasted until the end of the Islamic period. The repertory of forms is reduced to mainly cooking wares (both in late Roman and Islamic periods), leaving Gutiérrez Lloret to conclude: 'It appears that the ceramic forms that define this horizon [ceramics of the emiral and caliphal periods] have their origin in Late Roman prototypes (sixth and seventh centuries) in the Alicante region, more clearly related to North African production than to the properly Visigothic world of the interior.'[4] She concludes, in addition, that the economic crisis of the sixth and seventh centuries also had the effect of constraining technology. This explains why, in this rural zone, technological advances of the Roman period had to be reinvented in early Islamic times, as demonstrated by the progressive recuperation of hand-building with a turntable (a Roman technique that had fallen into disuse in late Roman times) and the kickwheel.[5] The continuity in pottery forms must be explained, clearly, by the persistence of Hispano-Roman settlement in the first centuries of Al-Andalus, a continuity also documented in the ceramics of Murcia, Almería, Málaga, and Granada.[6] The periodisation of pottery finds in the Christian north, in broad lines, is congruent with that of Paleo-andalusí wares: indistinction between late Roman and Visigothic forms, which last into the ninth century or later, with turntable-modelled wares predominating until at least the twelfth century.[7]

3 *Ibid.*, pp. 18, 33, 122, 124, 132.
4 *Ibid.*, p. 237.
5 *Ibid.*, p. 248.
6 For similar findings in rural Cuenca, where late Roman forms persisted through the ninth century and no datable Islamic pottery appears until the tenth, see Yasmina Alvarez Delgado, 'Cerámicas comunes con y sin decoración, siglo X. Arcávica (Cuenca)', *II CAME, II,* 403–12. See also the useful survey of families of late Roman pottery forms in early medieval excavations, in L. Caballero Zoreda, 'Pervivencia de elementos visigodos en la transición al mundo medieval. Planteamiento del tema', *III CAME,* I, 111–34, on pp. 127–9.
7 Pedro Matesanz Vera, 'La cerámica medieval cristiana en el norte (ss. IX–XIII): Nuevos datos para su estudio', *II CAME,* I, 245–60, on pp. 248–9, 257. See also Ramón Jarrega Domínguez, 'Notas sobre una forma cerámica. Aportación al

What exactly these wares represented in terms of the culture of Muwallad/s and Mozarab/s is, of course, speculative. Gutiérrez Lloret supposes that when Roman commercial networks dissolved in the western Mediterranean of the sixth century, what was left were pre-existing local centres of pottery production, with an elemental technology. These produced a reduced repertory of forms: kitchen receptacles with convex or flat bases, straight walls and wide mouths, together with a few other forms, including storage vessels, found equally in urban and rural places. In the Paleoandalusi period, such forms continued to be produced by Muwallad/s and Mozarab/s, but had become 'relics of an inoperant urban social organisation'.[8] A broadening and new standardisation of the repertory of forms or the introduction of glazing are generally taken as a reflection, or index, of the Islamisation of this population. Thus in Pechina, the emergence in the more recent of two archaeological levels of a standard caliphal repertory is taken as an index of cultural homogenisation through the Islamisation of diverse populations, including the original Hispano-Roman one.[9] In Motril, the turntable tradition lasts through the fall of the Caliphate and can be explained, according to Gómez Becerra, by the persistence in this region of human groups associated with this technique, communities of indigenous origin, even though characterised at these late dates by a degree of Islamisation, judging by the presence of new forms, especially glazed ones. From the tenth century on, there was a greater repertory of forms and decorations in turntable-produced wares. Gómez Becerra asks if the process of Islamisation in these communities didn't bring about a rapid change in the ceramic repertory.[10] Finally Manuel Acién Almansa, commenting on unglazed turntable wares from Bezmiliana, a repertory diffused throughout a wide area of south-eastern Al-Andalus, from Ibiza, to Alicante, Murcia, Almería, Granada, Málaga, and as far as Ceuta on the Moroccan coast, corresponds to 'the society that inherited the Hispano-Gothic world', though not *all* Hispano-Romans, simply

estudio de la transición del mundo romano al medieval en el este de Hispania', *I CAME*, II, 305–13.

8 Sonia Gutiérrez Lloret, 'La cerámica paleoandalusí del sureste península (Tudmir): Producción y distribución (siglos VII al X)', in Antonio Malpica Cuello, ed., *La cerámica altomedieval en el sur de Al-Andalus* (Granada, Universidad, 1993), pp. 37–65, on pp. 43–4, 49.

9 Francisco Castillo Galdeano and Rafael Martínez Madrid, 'Producciones cerámicas en Bajjana', *ibid.*, pp. 67–116, on p. 116.

10 Antonio Gómez Becerra, 'Cerámica a torneta rodente de "El Maraute" (Motril). Una primera aproximación a la cerámica altomedieval de la costa granadina', *ibid.*, pp. 172–91, on p. 191.

village communities associated with rural, particularly mountainous (*saltus*) habitats, whose remoteness explains the characteristic technical deficiencies of such pottery.[11]

Archaeologists therefore associate this repertory of wares with a habitat and thus with a specific social group or groups, presumed to be in possession of an 'indigenous' culture. But beyond such generalities, the problem is severely undertheorised. First, as Barceló noted in a trenchant comment at the meeting where the above data was presented, archaeologists prefer the term 'habitat' (as in 'hilltop habitat') because neither they nor historians have clarified what the nature of a village settlement was in the high Middle Ages, or even how to define a settlement. Archaeologists (as the debate over the social and cultural meaning of unglazed kitchenwares shows) have privileged the zone of residence over the zone of work.[12]

Then there is the problem of cultural boundaries. Cultural or ethnic boundaries ('degree of enclosure', anthropologists now say) are determined by a variety of factors, both material and ideological. To posit an array of material objects as an index of enclosure is not warranted unless the complex nature of enclosure is made explicit. When the Arabs and Berbers arrived in 711, acculturation began immediately and can be reckoned a two-way street. Now two or more repertories coexisted: but clearly such coexistence was not static. As Helena Kirchner observed, it is absurd to posit that Arabs introduced glazed wares while all unglazed kitchenwares are judged *ipso facto* 'indigenous'. Surely the Arabs didn't cook with their best wares![13]

The ability of 'indigenous' turntable or hand-building potters to adopt new forms brought by Arabs and Berbers and produce them with their accustomed technology can be appreciated from the case of the Arab portable oven, or *tannūr*. These ovens were modelled by hand or on the turntable and examples have been excavated at the Rābiṭa at Guardamar, at El Sompo (Cocentaina), and at Ermita de les Animes (Gandia), all ninth- and tenth-century sites in the current provinces of Valencia and Alicante. Its appearance in south-eastern Al-Andalus in this period makes it, in Gutiérrez Lloret's view, an effective indicator of the steady cultural Islamisation of indigenous populations.[14]

11 Manuel Acién Almansa, 'Cerámica islámica arcaica del sureste de Al-Andalus', *Boletín de Arqueología Medieval*, 3 (1989), 123–35, on pp. 134–5; and 'La cultura material de época emiral en el sur de Al-Andalus. Nuevas perspectivas', in Malpica Cuello, ed., *Cerámica altomedieval*, p. 170.

12 Barceló, discussion in Malpica Cuello, ed., *Cerámica altomedieval*, p. 203.

13 Helena Kirchner, discussion in Malpica Cuello, ed., *Cerámica altomedieval*, p. 146.

14 Sonia Gutiérrez Lloret, 'La producción de pan y aceite en ambientes domésticos. Límites y posibilidades de una aproximación etnoarqueológica', Formas de habitar e alimentação na Idade Media (Mértola, 17–20 September 1993), typescript.

The analogical value of such a notion of a Paleoandalusi material culture, based on a few well-documented phenomena (a limited repertory of unglazed cookwares with late Roman forms made on a turntable or hand-built), with well-defined chronological parameters (from the sixth to the ninth centuries) is nevertheless clear: it suggests a framework for the interpretation of Hispano-Roman culture and society and the elucidation of the processes of cultural change as the society evolved from a Latin-speaking Christian one to a normative, Arabic-speaking, Muslim one. To the historian, a *terminus ad quem* of the mid-ninth century is also suggestive. The register of material culture renders palpable the cultural change that Arab historians recount as having taken place during the reign of the Emir ᶜAbd al-Raḥmān II (822–852), whose adoption of ᶜAbāssid administrative and social norms marked the evolution of a garrison state ruling a Christian majority into a more homogeneously Islamic polity.[15]

Juan Zozaya, whose research has included pottery, but also work on castles, burial sites, and extensive archaeology surveys, has generated a minute table of periodisation whose validation would vastly increase the task before us.[16]

711–756	Preandalusi
756–852	Paleoandalusi
852–876	Protoandalusi
876–925	Pre-Umayyad
925–944	Proto-Umayyad
944–1000	Umayyad
1000–1035	Epi-Umayyad
1035–1086	Post-Umayyad

These 'phases' in turn are broken down into sub-phases which are not so much chronological, but stylistic. Thus the Preandalusi phase includes Hispano-Roman, Visigothic, Byzantine, and Paleoberber sub-phases which relate to specific features of the material record. Rather than so many small phases, I would prefer a longer and less differentiated Paleoandalusi period lasting, for reasons we will presently examine, until the end of the *fitna* of 879–927, which put

15 On ᶜAbd al-Raḥmān II's reign as a social, political and cultural turning point, see my *Islamic and Christian Spain in the Early Middle Ages: Comparative Perspectives on Social and Cultural Formation* (Princeton, Princeton University Press, 1979), pp. 200–4.

16 In personal communications, Gutiérrez Lloret is extremely hesitant to extend the compass of her 'paleoandalusí' period to the culture at large, while Zozaya displays no such reticence.

an end to an older social organisation associated with the indigenous population.

Here, I will comment only on his construction of Paleoandalusi culture. He first factors into three phases the period from the Arab conquest to the end of Muḥammad's reign (711–876): a Preandalusi phase corresponding to the period of the conquest up to the instauration of Umayyad rule (711–756), the Paleoandalusi phase proper, bracketed by the reigns of the first two ᶜAbd al-Raḥmān/s (756–852), and a third, short Protoandalusi phase embracing the reign of Muḥammad. These phases had sub-periods which relate to specific, characteristic artistic styles and which need not detain us here. My conclusion is that Zozaya's phases are too closely linked to standard political chronology to merit any distinction between Pre-, Paleo-, and Proto-, and that a more inclusive Paleoandalusi period embracing the entire period 711–876 provides a framework which is not only much more flexible, but which also permits the fashioning of a distinctive sociocultural configuration quite different from that which followed it.

Zozaya, however, provides a number of elements in addition to pottery which contribute to a fuller portrait of Paleoandalusi culture (including Zozaya's pre- and proto-phases as well).[17] These include (but are not limited to):

(1) the sharing of graveyards between Muslims and Christians (for example, the Visigothic cemetery of Segóbriga (Cuenca) antedates 711 but continued thereafter. Along with many Christians are buried twenty-three Muslims, identified by skeletons resting on their right side, with the head towards the south, facing east (the direction of Mecca); these Muslims were buried with some of their possessions, in Visigothic fashion.[18]);

(2) the sharing of churches;

(3) bilingual coinage;

(4) a *commendatus* style of defensive settlements (state support of tribal units of settlement, that is, a mixed state and tribal initiative);

(5) continuity of late Roman pottery types, yielding around AD 825 to glazed wares with underglaze incised decoration in the Andalusi heartland, but not elsewhere;

17 Juan Zozaya, 'The Islamic Consolidation in Al-Andalus (8th–10th Centuries): An Archaeological Perspective' (typescript), and 'Importaciones casuales en Al-Andalus: Las vías de comercio' (typescript).

18 Manuel Retuerce Velasco and Alberto Canto García, 'Apuntes sobre la cerámica emiral a partir de dos piezas fechadas por monedas', *II CAME*, III, 93–104. See Caballero Zoreda's summary of transitional necropoli, 'Pervivencia de elementos visigodos', pp. 126–7.

(6) the defensive system of watch-towers (*atalayas*), begun early and completed after the reign of ᶜAbd al-Raḥmān I.

Paleoandalusi settlements: the marshes of the eastern coast[19]

On the eastern Mediterranean littoral there are ample zones of marshland known generically by the Arabism *marjal*. These are places with high water tables, where settlement is possible on hillocks rising above the marsh or on inland hills identified by the toponym *cabezo*. Their settlement in the eighth and ninth centuries represented a replacement of late Roman sites at higher altitudes by *cabezos* abutting or within marshland where a distinctive agricultural economy could be practised. This economy was based on the rich biomass of littoral marshlands which supply not only a variety of edible seafoods, but also fibres and canes that can be used in household crafts, and an agriculture that takes advantage of seasonal flooding. Here too, the earliest irrigation systems were implanted, before the famous huertas of the lower Segura were developed, in the lower sector of the river formed by a set of interlaced canals and meanders, rather than by a defined river bed. Al-ᶜUdhrī, describing this area, states that the Segura's channel ends to the south of al-Qaṭrullāt 'in a district called al-Muwalladūn in the direction of the *alquería* called al-Juzaira'. The passage makes clear that this was an area of Muwallad settlement. The pottery register displays acculturation in process through the appearance of new pottery forms, not contained in the late Roman repertory: *tannūr*/s (as noted above), *jarros*, and noria pots (*qawādīs*; singular, *qādūs*). Norias are useful in waterlogged districts because irrigating with them also lowers the water table. *Qādūs* sherds are found in all of these settlements, the earliest from the mid-eighth century at Cabezo del Molino (Alicante).[20] Map 5 depicts settlement

19 The following discussion is based on Sonia Gutiérrez Lloret, 'Espacio y poblamiento paleoandalusí en el sur de Alicante: Orígen y distribución', *III CAME*, II, 341–5; Sonia Gutiérrez Lloret, 'La formación de Tudmir desde la periferia del estado islámico', *Cuadernos de Madīnat al Zahrā*, III (in press). Sonia Gutiérrez Lloret, 'El tránsito de la antigüedad tardía al mundo islámico en la cora de Tudmir: Cultural material y poblamiento paleoandalusí', unpublished doctoral dissertation, University of Alicante, 1992; Rafael Azuar Ruiz and Sonia Gutiérrez Lloret, 'Formación y transformación de un espacio agrícola islámico en el sur del País Valenciano: El Bajo Segura (siglos IX–XIII)', *Castrum 5*. For a similar kind of exploitation of marshland in medieval Sicily, see Henri Bresc, 'Le eaux siciliennes: une domestication inachevée du XIIe au XVe siècle', in Elizabeth Crouzet-Pavan and Jean-Claude Maire-Vigueur, eds., *Water Control in Western Europe, Twelfth–Sixteenth Centuries* (Milan, Università Bocconi, 1994; Proceedings, Eleventh International Economic History Congress, vol. B2), pp. 73–85, on pp. 74–80.

20 Sonia Gutiérrez Lloret, 'El origen de la huerta de Orihuela entre los siglos VII y XI: Una propuesta arqueológica sobre la explotación de las zonas húmedas del

in the lower Segura between the eighth and tenth centuries, characterised by settlements either in the marshland or on hillsides. Map 6, locating settlements of the eleventh to the thirteenth centuries in the same zone, shows the development of the huerta network of irrigation canals and the displacement of settlement from hillsides to the borders of the huerta.

In Oliva, whose huerta probably had a similar history, a medieval noria site has yielded 5,000 *qādūs* fragments, the earliest from the first half of the tenth century.[21] Further to the north, the *marjals* of the northernmost sector of the huerta of Valencia were the site of early irrigation from wells. As in Murcia/Orihuela, the littoral zone no doubt was developed first. The placement of villages in the lower reaches of the Moncada irrigation canal, right on the perimeter of the marshland, on the site of the future huerta, is similar – if not identical – to the settlement pattern in the lower Segura depicted in Maps 5 and 6.[22]

The same kind of sequence describes the development of irrigation in the Ribera del Júcar, which, before the Acequia Real was built in the thirteenth century, had under Muslim rule been a flood plain irrigated either by norias or from affluents of the Júcar, but not from the Júcar itself.[23] When the Acequia Real was built, previously irrigated *alquerías* like Alásquer, Cabañés, Mulata, and Resalany – all along the Riu dels Ulls – were deserted.[24] Alásquer means 'military encampment' in Arabic and the place-name probably dates back to the eighth century.[25] Resalany, or Rasalany, means 'head of the spring' (*ra's al-ᶜayn*), indicating irrigation from a spring rather than a river.

Bajo Segura', *Arbor*, 1995, p. 83: the Cabezo del Molino pots are similar to Egyptian forms and were likely introduced by Egyptian soldiers (*jundī*/s) settled in the area in the eighth century.

21 André Bazzana, Salvador Climent and Yves Montmessin, *El yacimiento medieval de 'Les Jovades' – Oliva (Valencia)* (Gandia, Ayuntamiento, 1987).

22 See Vicente Sales Martínez, 'La cuestión del extremal en el regadío de la Real Acequia de Montcada', *Cuadernos de Geografía* (Valencia), 44 (1988), 221–34, map on p. 229.

23 Joan F. Mateu Bellés, 'Assuts i vores fluvials regades al País Valencià', *in Los paisajes de agua. Libro jubilar dedicado al profesor Antonio López Gómez* (Valencia/Alicante, 1989), pp. 165–85, on pp. 171–4.

24 The pattern is obvious from Antonio José de Cavanilles's map of the Ribera in *Observaciones sobre el Reyno de Valencia*, 2 vols. (Zaragoza, CSIC, 1958), opposite p. 272, where the *despoblados* are marked with a cross. I am indebted to Joan Mateu for pointing this out to me.

25 On the distribution of toponyms from *al-'askar* in the peninsula, see Joaquín Vallvé, *Nuevas ideas sobre la conquista árabe de España: Toponimia y onomástica* (Madrid, Real Academia de la Historia, 1989), pp. 107–11.

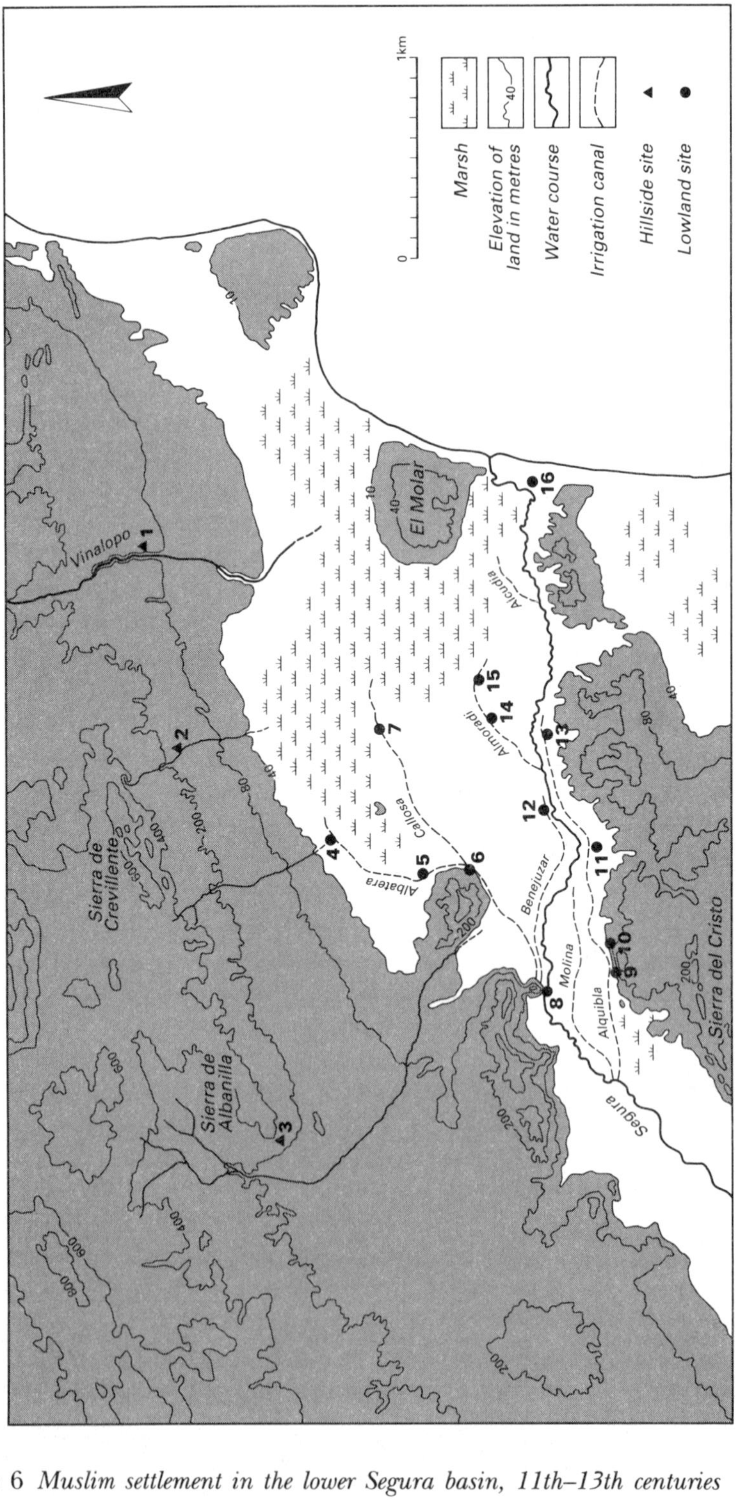

Map 6 *Muslim settlement in the lower Segura basin, 11th–13th centuries*

Paleoandalusi high culture: science

Here I will suggest that, to the extent that the notion of a Paleoandalusi culture is a useful concept for describing a material culture in the process of formation, the same can be said of high culture. In the area of science, astronomy/astrology in particular, early Andalusi science coexisted in a kind of symbiosis with a very humble Low Latin science, diffused by Mozarabs and which still showed signs of life in the eleventh century, in the description of Julio Samsó.[26] The heart of Samsó's notion is that until the first Eastern *zījāt* – astronomical tables – reached Al-Andalus in the reign of ᶜAbd al-Raḥmān II,[27] Andalusi astronomers and astrologers had to use astronomical methods already in use by Christians. Thus we know that the so-called *Book of the Crosses* was in use in Córdoba at the end of the eighth century when al-Ḍabbī cast horoscopes during the reign of the Emir Hishām I. This book was most likely translated into Arabic, possibly by al-Ḍabbī, from a lost Latin original, revised in the eleventh century by ᶜUbayd Allāh al-Istijī and then translated into Castilian at the order of Alfonso X the Wise. Using this method, astrologers could cast horoscopes without the use of astronomical tables simply by ascertaining in which sign of the zodiac Saturn, Jupiter, Mars, and the Moon were.[28]

For the purposes of calendrical calculation – as in the famous *Kitāb al-Anwā'*, or *Calendar of Córdoba*, a text of the tenth century – three sources are mentioned: the *Sind-Hind*, that is the Indian *zij* brought to the peninsula in al-Khwārizmī's recension in the mid-ninth century; the 'method of the observers', a reference to tables based directly on the observation of their author, in this case al-Battānī; and finally the 'calculators', namely Hippocrates and Galen. The allusion has to do with the Greek tradition of dividing the seasons unevenly: four months each for winter and summer, two each for spring and autumn. Hippocrates used a 'natural' calendar

26 Julio Samsó, 'En torno a los métodos de cálculo utilizados por los astrólogos andalusíes a fines del s. VIII y principios del IX: Algunas hipótesis de trabajo', *Actas de las II Jornadas de Cultura Arabe e Islámica (1980)* (Madrid, Instituto Hispano-Arabe de Cultura, 1985), pp. 509–22, on p. 520, and Julio Samsó, *Las ciencias de los antiguos en Al-Andalus* (Madrid, Mapfre, 1992), Chapter 1: 'Arabic Science and Isidorean Culture'.

27 Luis Molina, *Una descripción anónima de Al-Andalus*, 2 vols. (Madrid, CSIC, 1983), I, 138 (text); II, 146 (trans).

28 Julio Samsó, 'The Early Development of Astrology in Al-Andalus', *Journal for the History of Arabic Science*, 3 (1979), 228–43. See also Samsó's edition of that part of the *urjūza* (poem) of Al-Ḍabbī that appears, Castilianised, as Chapter 57 of the Alfonsine *Libro de las Cruces*, 'La primitiva versión árabe del Libro de las Cruces', in Juan Vernet, ed., *Nuevos estudios sobre astronomía española en el siglo de Alfonso X* (Barcelona, Institución Milá i Fontanals, 1983), pp. 149–61.

based not on abstract divisions but on natural phenomena such as climatic phenomena or the arrival and departure of migratory birds. The *Calendar of Córdoba* reflects this tradition when it recommends that shipping cease on 17 November, reflecting Hesiod's recommendation that navigation cease with the rising of the Pleiades. Likewise the *Calendar* provides dietary rules, by seasons rather than months, and strictly in accordance with the norms of humoral pathology established by Hippocrates and Galen.[29] Furthermore, as Willy Hartner explained, calendars of the *anwā'* type, that is, those based on risings and settings of specified stars or constellations, represented an ancient way of establishing the solar calendar, which all Near Eastern peoples used for agriculture. The *anwā'* formed a kind of 'solar skeleton' on to which the lunar, religious year was then fitted.[30]

To the same tradition belongs the Arabic reflection of the *Epistula metrica* of the Visigothic King Sisebut, a treatise containing an explanation of solar and lunar eclipses sent by the King to Isidore of Seville. In the *Crónica del moro Rasis*, the King is described as having in his court 'a great philosopher and very good astrologer named Çirdan', Çirdan being a deformation, via Arabic, of Isidore.[31]

We can also define a Paleoandalusi medicine, dominated by Christians until the mid-tenth century. In his book on *The Generations of Physicians*, the *muwallad* Ibn Juljul enumerates medical doctors practising during the reigns of Muḥammad and ᶜAbd al-Raḥmān II. These included a monk named Jawād, famed for his medicines and potions; Khālid ibn Yazīd ibn Rumān, similarly reputed for his botanical drugs; and Ibn Malūka, a Cordoban surgeon and bleeder who had thirty chairs for patients set up at the door of his house. Ibn Juljul also recounts how an old monk cured ᶜAbd al-Raḥmān III of an earache. Ibn Juljul declares that he himself used Latin sources which included the *Etymologies* of Isidore of Seville, Orosius, and a Latin version of a Greek chronicle by Eusebius, bishop of Caesaria, translated into Arabic in tenth-century Córdoba. Samsó concludes that two distinct kinds of medicine coexisted in Al-Andalus well into the tenth century: 'a popular medicine of Latin descent and a more cultivated medicine practised by physicians

29 Julio Samsó, 'La tradición clásica en los calendarios agrícolas hispanoárabes y norteafricanos', *Second International Congress of Studies on Cultures of the Western Mediterranean* (Barcelona, n.p., 1978), pp. 177–86.

30 Willy Hartner, Discussion, *Oriente e Occidente nel Medioevo: Filosofia e Scienza* (Rome, Accademia Nazionale dei Lincei, 1971), pp. 544–5.

31 Julio Samsó, 'Nota sobre la biografía del rey Sisebuto en un texto árabe anónimo', *Serta gratulatoria in honorem Juan Régulo*, I Filología (La Laguna, Universidad, 1985), pp. 639–42.

who had access to the Greek works of Hippocrates and Galen and which had been translated into Arabic in the East and introduced in Al-Andalus'.[32]

Geographical knowledge presents problems of a different order because here we are concerned not only with textual traditions but with a broader process of *iqlīmatiyya al-Islām*, whereby the Muslims had to, as it were, 'solve the landscape' of their new home, a multifaceted process which included Arabising Latin place-names, or replacing them with semantic calques, which in the case of interesting or important toponyms frequently involved the creation of stories to explain the meaning of the names, sometimes drawing on Latin geographical and historical traditions. Those historical traditions also found a learned expression as useful Latin materials were translated or otherwise ingested into the nascent corpus of Arabic geographical and historical lore that became the basis for the creation of a more formal Andalusi history and geography. This was a process which was activated in the eighth century and therefore forms part of Paleoandalusi culture.

Vallvé provides interesting discussions of many of the place-names associated with the Arab conquest. Thus, to give one particularly complex example, the Arabs named an island off Cádiz Umm Ḥakīm in honour of a slave of the conqueror Ṭāriq ibn Ziyād, or so the story goes. But Umm Ḥakīm is merely a literal rendering in Arabic of the isle of Hera or Juno, whose Latin name – Insula Junonis – is perpetuated in the current place-name Isla de León.[33] Thus for Vallvé the intense Romanisation of the area first conquered meant that Latin culture (both the place-names themselves and the traditions associated with them) would continue to be reflected. The first descriptions of these places (Gibraltar, Tarifa, Algeciras, and so forth) in eastern Arab accounts are all vague and contradictory – descriptions taken from Greek authors, with their corresponding translations and glosses. All the place-names around the Straits of Gibraltar which became invested with a mythical quality surrounding the events of the conquest, as reported in later accounts, were popular etymologies explained by reference to real or fictitious persons (e.g., Jabal Ṭarīq to Ṭarīq ibn Ziyād, Ṭarīfa to a putative

32 Juan Vernet, 'Los médicos andaluces en el "Libro de las Generaciones de Médicos", de Ibn Juljul', in *Estudios sobre Historia de la Ciencia Medieval* (Barcelona, Universidad, 1979), pp. 469–86; Julio Samsó, 'Astrology, Pre-Islamic Spain and the Conquest of Al-Andalus', *Revista del Instituto Egipcio de Estudios Islámicos en Madrid*, 23 (1985–86), 79–94.

33 Vallvé, *Conquista árabe*, p. 93. Jabal Ṭarīq (Gibraltar) itself he believes to be an Arabisation or translation of a Roman or pre-Roman toponym, probably Calpe (*ibid.*, pp. 60, 120).

conqueror named Ṭarīf, and so forth). Arab authors repeatedly translated and glossed a small set of texts, the most important of which was the *Historiae adversum paganos* of Orosius which is where, for example, they found a cape (Arabic *ṭarf*) which became Ṭarīfa, for which an eponymous Ṭarīf had to be invented.[34]

This process was neither consistent nor did it obey any inherent logic, which is why the origin of many such place-names have presented such conundrums to historians and philologists. It is an example of the process by which Arabs, colliding with a civilisation of late antiquity, shaped a new civilisation (and history) out of old materials, which they integrated piecemeal and without much effort to understand the nature of past civilisations in any but an *ad hoc* way: the 'debris of an obliterated past', in Crone's characterisation.[35]

Orosius was the key source for the history of Al-Andalus before the conquest and was known through a mid-tenth century translation made by Qāsim ibn Aṣbag.[36] However this 'Hurūshyūsh' was an interpolated, not a literal, translation, and included Christian sources known previously to Andalusi scholars, including Isidore's *Historia Gothorum* and most likely a *Book of the Prophets*, a lost Mozarab history written after the Muslim conquest. The use to which later Arab geographers put the *Hurūshyūsh* can be appreciated from the permutations of Orosius's description of Spain as a triangle, an example of the process which I have described as 'solving the landscape' by interpreting Christian documents.[37] Although the *Hurūshyūsh* is somewhat later than the Paleoandalusi period proper, it is a product of the characteristic cultural processes of that period, 'a reflection of the cultural situation of al-Andalus in a moment in which the poor vestiges of the classical world are appropriated by the flourishing Hispano-Muslim culture'.[38] When Sāᶜid al-Andalusī wrote his history of science in the eleventh century, he made no mention of Isidore or Orosius and his account of the 'Science of the Romans' deals mainly with eastern Christians in the court of the ᶜAbbāsids.[39]

34 Following Vallvé, *Conquista árabe*, pp. 120–1.

35 Patricia Crone and Michael Cook, *Hagarism: The Making of the Islamic World* (Cambridge, Cambridge University Press, 1977), p. 73; Patricia Crone, *Slaves on Horses: The Evolution of the Islamic Polity* (Cambridge, Cambridge University Press, 1980), p. 10.

36 On the influence of Orosius on Arab historians and geographers, see Joaquín Vallvé, 'Fuentes latinas de los geógrafos árabes', *Al-Andalus*, 32 (1967), 241–60; Diego Catalán and María Soledad de Andrés, eds., *Crónica del Moro Rasis* (Madrid, Gredos, 1974), pp. xl–lxi; and Luis Molina, 'Orosio y los geógrafos hispanomusulmanes', *Al-Qanṭara*, 5 (1984), 63–92.

37 See texts arranged for comparison by Molina, 'Orosio', pp. 72–80.

38 *Ibid.*, p. 92.

39 Sā'id al-Andalusī, *Science in the Medieval World: Book of the Categories of Nations*, ed. S. I. Salem and A. Kumar (Austin, University of Texas Press, 1991).

The formation of an Andalusi vernacular variant of Arabic, in contact of course with an initially majoritarian Romance-speaking population, is another Paleoandalusi phenomenon. Corriente denotes three strata of linguistic interference of Romance on the emerging Andalusi vernacular, differentiated chronologically.[40] The first stratum is when Romance is a substratum, 'a witness to the process of the substitution of one language by another in the population during the first phase of its Arabisation', in the first several generations after the conquest. Common Romance terms for which no Arabic equivalent existed were simply adapted into vernacular Arabic, 'with the necessary morpho-phonemic adaptations'. Examples are *lúp*, wolf (Romance, *lupum*), *istípa*, steppe (the latter, an example of another part of the process of 'solving the landscape'; I might add that all of the early toponyms mentioned above belong in this category). In other cases, for social reasons related to the daily interaction of the two different linguistic populations such as, for example, the typical phenomenon of bilingual children born to Romance-speaking mothers, certain Romance words were adopted into the Andalusi vernacular in spite of the existence of standard Arabic equivalents (Corriente gives as examples a number of vulgar terms related to sexual organs).

The next group are 'adstratic', referring to a chronologically later, but still Paleoandalusi by my definition, period, after Andalusi Arabic 'had acquired its own personality and was used by broad sectors of the population, with or without active bilingualism'. This involved a second wave of Mozarabic loan-words, mainly rural, where the pace of acculturation was slower. Here, adaptation to a new environment produced numerous variations of Romance animal, plant, geographical, and agricultural terms (e.g., *aghrīl*, Spanish *grillo*, cricket; *duntal*, Spanish *dental*, ploughshare). The final group are 'suprastratic', Romancisms adopted from northern Christian usage, and which need not concern us here because this process was proper to a later period.

Conversion to Islam

It is clear that the dynamics of acculturation considered here in a variety of cultural areas, from pottery to ideas to language, were closely tied to the overall process of conversion. Here, a number of discrete but related issues are in play: what was the rate of acculturation? Were there differential rates for Islamisation and

40 Federico Corriente, *Arabe andalusí y lenguas romances* (Madrid, Mapfre, 1992), pp. 132–42.

Arabisation? Were there differential rates of conversion in different parts of the country? What variables affect acculturation rates? Are there cultural markers from the material (archaeological) record that can be related to the curve of conversion?

In Richard Bulliet's account of conversion to Islam, the rate of growth of the convert population follows a logarithmic curve and works by contagion: the more Muslims, the greater the probability of acculturative contact between the unconverted population and Muslims.[41] The curve of conversion accounts for numerous political events of the early Islamic centuries because political appeals appropriate to a mainly unconverted population were quite different from those appropriate to a majoritarian Muslim population. At the moment of the most rapid increase, when the conversion curve enters its period of logistic growth, political and social life were highly volatile and could change abruptly in a period of only a few decades. Thus in Al-Andalus, the period just before the explosive period of growth was marked by the *fitna* of the late ninth century and the revolt of Ibn Ḥafṣūn, while that right after the logistic period had been completed was characterised by the fall of the Caliphate.

Bulliet divides the curve into stages determined by the progress of conversion among the indigenous population. The first 2.5 per cent are innovators, the next 13.5 per cent, early adopters, the next 34 per cent early majority, the next 34 per cent late majority, and the last 16 per cent laggards. *All the phenomena here described under the general rubric 'Paleoandalusi' are early adopter and, particularly, early majority phenomena.* In Bulliet's curve, the early adopter phase ends in 816, while the early majority phase extends from 816 to 961. In 816, only 16 per cent of the indigenous population would have converted, by 961, 50 per cent. In Bulliet's telling, these stages are significant because the texture of Arab polities changes radically once Muslims outnumber the indigenous, unconverted population. Before, there is no need for the rulers to develop mass social institutions. The Arab tribal structure suffices for the rule of a garrison state and specifically Muslim institutions are not required. When the State becomes predominantly Muslim, it gains *ipso facto* the psychological and social basis for declaring its independence, no longer requiring the security of belonging to an imperial structure located in the East. The same stage of conversion (coinciding with

41 Richard Bulliet, *Conversion to Islam in the Medieval Period: An Essay in Quantitative History* (Cambridge, MA Harvard University Press, 1979; Richard Bulliet, 'Conversion to Islam and the Emergence of a Muslim Society in Iran', in Nehemia Levtzion, ed., *Conversion to Islam* (New York, Holmes and Meier, 1979), pp. 38–51. See also Glick, *Islamic and Christian Spain in the Early Middle Ages*, pp. 33–5.

the end of the 'early majority' stage) saw revolts by non-Muslims or recent converts in many Islamic societies, including rebellions by Copts in Egypt and Zoroastrians in Persia. Ibn Ḥafṣūn's revolt, with substantial Mozarab support, satisfies the curve's prediction.[42] All of the institutions of mass conversion such as *madrasa*/s (i.e., the Islamic educational system), Sufi Orders, guilds, and so forth, only became significant in the eleventh century. Therefore the Paleoandalusi institutions, social movements, and reflections in material culture were quite unrelated to an *Islamic* society *per se*, but rather had to do with a garrison state in which Arab elites set the social, political, and cultural tone, but in which the unconverted and recently converted population continued to use older social and cultural forms, subjected to a variety of pressures from the Arabs. Therefore, the pace of acculturation can be presumed to have been *slow*, because religious conversion alone did not necessarily stimulate cultural borrowing until the advent of a late majority state.

However, significant changes were registered in the early majority period when the Mozarab family came under direct attack, when ᶜAbd al-Raḥmān II took steps to easternise, if not Islamise, the administration of the State, as well as the social norms it promoted.

There is an interesting counterpoint among historians of Al-Andalus with regard to the rate of conversion. Acién Almansa, on the one hand, posits a late conversion for the east of the country (Sharq al-Andalus), one squaring with the Bulliet curve; the charge to the Umayyad governor and prince, al-Balansī, in the early ninth century was to administer a no-man's land. Segmentary organisation was weak: one does not find the typical southern pattern of the 'aristocratisation' of local clans. In the eleventh century, the absence of clearly differentiated clans permitted the creation of the so-called Slav (*ṣaqāliba*) Taifas in the region, rather than states organised through Arab or Berber lineages. Thus Islamisation for Acién is not a lineal process, but rather a complex one dependent on tribalism, on the 'feudalised' social forms of indigenous society (who, in their villages, come to approximate a tribal society); on aristocratisation in tribal circles producing feudal-like behaviours, and so forth. Finally, for Acién, the State is the principal motor of Islamisation: as it makes its control of the countryside more effective it provides both the means and agencies for religious conversion.[43] Acién Almansa's viewpoint is important and I will return to it shortly when discussing the indigenous response to conquest.

42 Bulliet, *Conversion to Islam*, p. 125.

43 Manuel Acién Almansa, 'Sobre la función de los *ḥuṣūn* en el sur de Al-Andalus. La fortificación en el Califato', in *Coloquio Hispano-Italiano de Arqueología Medieval*

In the case of conversion to Islam in the Valencian region, Epalza, in an influential article written with the classical archaeologist Enrique Llobregat, argues that there were few Christians there before 711; the region had only been lightly Christianised in the first place, and Christians as such, just like their brethren in North Africa, disappeared by the second generation after the Muslim conquest, through lack of a bishopric. The Muslims would not have recognised such communities as Christian, lacking any religious chief or organisation. They criticise Bulliet's methodology for not giving sufficient importance to specific juridical mechanisms on which conversion was based; Bulliet assumes the prevalence of norms more suited for a society in which free and personal conversion is recognised. Conversion in the Islamic world was not an individual or personal phenomenon but a civic and legal one. Therefore, the moment that Visigothic governance was replaced by a Muslim one, pagans converted *en masse* to avoid being killed. There were individual Christians in Valencia, of course, but lacking an organised community, they would have been considered officially Muslim. Because no sacraments were available, the weakening of belief and concomitant Islamisation would have been completed by around 800. Thus, in Valencia, there were no 'Mozarabs'.[44]

The issue of the Valencian 'Mozarabs' is a highly polemical one, doubly so because any stance for or against their existence marks the author politically. The regional right, particularly during the Franco period, argued the distinctiveness of Valencian language and culture by asserting that the region's language antedated the Aragonese and Catalan conquest. The indigenous population, whether Muslim or Christian, spoke the same romance language as the Valencian 'Mozarabs' had; therefore Valencian language and culture is not Catalan. To assert there were no Mozarabs, then, is to attack the regionalist cultural hypothesis directly.[45]

Be that as it may, there are problems with the Epalza–Llobregat hypothesis. In the first place, Bulliet did not invent diffusion theory. Innovations are adopted by 'contagion', no matter what the pre-

---

(Granada, Patronato de la Alhambra, 1992), pp. 263–4. I can accept the view of the State's instrumentality in conversion of Muwallad/s, but only in a non-formal sense: the more the power of the State is felt, the more will social (especially prestige) and economic factors weigh in inducing individuals to convert. The State was not in the conversion business.

44 Míkel de Epalza and Enrique Llobregat, '¿Hubo mozárabes en tierras valencianas? Proceso de islamización del levante de la península (Sharq al-Andalus)', *Revista de Estudios Alicantinos*, 36 (1982), 7–31.

45 The regionalist position is intelligently defended by Leopoldo Peñarroja Torrejón, *Cristianos bajo el Islam: Los Mozárabes hasta la reconquista de Valencia* (Madrid, Gredos, 1993).

vailing institutional or socioeconomic structure might have been. Second, it is doubtful that the Muslims of the eighth century would have been able to discriminate among shades of belief or non-belief of the subject population: it was well-known that Spain was a 'Christian' country. Third, in the Islamic East the norms regarding assimilation of non-Arab subjects crystallised by the early eighth century, in a doctrine of clientage (*walā'*) whereby non-Arabs who wished to convert were obliged to find an Arab patron, whose client (*mawla*) he became. In this fashion, Berbers, Hispano-Romans and others adopted Arab tribal names and clientage was suffused with what Crone calls a 'tribal after-image'.[46] This kind of fictive kinship becomes extremely significant in understanding the social situation of Muwallad/s. Although in tribal societies it may have been common for entire segments to convert *en masse* following their *shaykh*/s, in no way can one envision mass conversion among detribalised Hispano-Romans. Conversion in Spain could only have been individual (barring the exception, limited to the conquest period of 711–16 only, of local populations who resisted surrender). Bulliet's rules therefore apply to the entire peninsula, and Valencia cannot be excepted.

Epalza presumes, incidentally, that cultural and linguistic Arabisation proceeded at a much slower pace than did conversion, inverting the logic of Bulliet's argument. He enumerates several political and social factors, including ᶜAbd al-Raḥmān's victory over dissident Muwalladūn and popular anti-Fatimid fervour of the tenth century as contributing to acculturation.[47] But such a construction ignores the dynamics of Bulliet's curve of conversion: by the early tenth century the conversion process would have, in any case, entered its explosive phase, providing an internal dynamic favouring acculturation. That dynamic is sufficient to explain pressure for assimilation experienced by remaining Mozarabs, a heightened sense of belonging to the Arab world on the part of Muwalladūn, and popular religious fervour directed against sectarians like the Fatimids.

Nor can the notion that Christian life requires an organised Church be sustained. Medieval Christians could be baptised without a bishop or even a priest, particularly in remote areas where laymen had to fill in.[48] As Bulliet observes,

46 Patricia Crone, *Roman, Provincial and Islamic Law: The Origins of the Islamic Patronate* (Cambridge, Cambridge University Press, 1987), especially on pp. 36, 91.

47 Míkel de Epalza, 'Relacions dels països catalans amb el món musulmà', *Revista de Catalunya* (February 1987), pp. 49–62, on pp. 54–5.

48 See, on this point, *Dictionnaire de Thèologie Catholique*, 2/1: 284–6; *New Catholic Encyclopedia*, 2:65, and Gratian, *Decretum*, pt. 3, *De consecratione*, IV, especially cols. 1367–82. E.g.: 'Etiam laici necessitate cogente baptizare possunt' (Ch. 21, col. 1368). I am indebted to Thomas Burman for these indications.

the argument based on the absence of bishops seems to me to be very weak indeed. If so, it may well indicate a Christian community with little religious leadership. But to jump from this to the idea that *therefore* everyone converted en masse to Islam is bizarre. Islamisation implies access to religious knowledge, integration into a community, acceptance of new types of religious leaders, change of social habits, and many other things. It is not simply a residual category people fall into because there is no bishop around.[49]

Guichard's discussion of Islamisation is more diffuse. He supposes the *rapid* Islamisation of indigenous peoples 'according to modalities that escape us completely'.[50] But he then goes on to assert, following the Bulliet thesis, that three-quarters of the population were Muslim by the end of the tenth century.[51] Later on in the same work, however, he cites figures supplied by Miquel Barceló to the effect that 560 of 773 *qurā* in the region of Córdoba were paying the Islamic tithe or *ʿushr* in the mid-ninth century, indicating that in that sample 72.44 per cent of the population had converted to Islam.[52] This pace is what Bulliet and I would characterise as slow, particularly when opposed to Epalza and Llobregat's *terminus ad quem* of AD 800 for Valencia. Guichard also notes that there is some evidence that, in Valencia, urban people had tribal *nisba*/s, while document witness-lists from the twelfth and thirteenth centuries show many rural Muslims with no tribal *nisba*/s. (The *nisba* was that part of the Arab name that indicates tribe, place of origin, or profession.) The logic of the concentration of tribal *nisba*/s in certain locales and not in others *may* respond to Arab settlement and the subsequent diffusion of that *nisba* among the local (perhaps Muwallad) population.[53] Guichard, oddly enough, does not use terms like 'client' or 'clientage', but clearly that is the issue at stake when one refers to the diffusion of tribal *nisba*/s among Muwalladūn. Indeed, the prevalence of rural Valencian Muslims

49 Personal communication, August 1993.

50 Guichard, *Musulmans de Valence*, I, 176.

51 Elsewhere, Guichard observes that qualitative data from places like Mérida and Toledo seem to indicate a very fast pace of Arabisation and Islamisation there; 'Les Mozarabes de Valence et d'al-Andalus entre l'histoire et le mythe', *Revue de l'Occident Musulman et de la Méditerranée*, 40 (1985), 17–27, on p. 24.

52 Pierre Guichard, *Les Musulmans de Valence et la Reconquête (XIe–XIIIe siècles)*, 2 vols. (Damascus, Institut Français de Damas, 1990–91), vol. II, 249, note 6.

53 *Ibid.*, I, 173–4. In his earlier volume (*Al-Andalus: Estructura antropológica de una sociedad islámica en Occidente* (Barcelona, Barral, 1976), p. 490), Guichard asserts that the number of Muslims with tribal *nisba*/s was still small in the caliphal era, but two centuries later, two-thirds of ibn al-Abbār's biographees had them. This is a reflection of the process of Arabisation, at a time when there was no longer any tribal reality.

with no tribal *nisba*/s might well support the Epalza–Llobregat thesis. But here, it is difficult to posit an eighth-century dynamic on the basis of much later documentation. It is doubtful that Arab and Berber settlements dating to the conquest were still segmenting by the twelfth century, meaning that very late converts may well not have adopted tribal *nisba*/s or have been *mawla*/s in the usual sense.[54]

Peter Scales, in his discussion of Berber settlement, is one of the few authors that uses Bulliet's curve to illustrate or corroborate social or cultural trends. The curve, he asserts, well accords with Federico Corriente's conclusions concerning the Arabisation of Berber lexical roots: namely, Berber influences are restricted to the rural sphere, whereas very few Berberisms turn up in urban speech, 'probably revealing the rapid assimilation of Berbers within an urban environment'.[55] However, this distribution of Berber linguistic inflections is just what one would expect from Guichard's picture of Berber settlement. If most Berbers were rural, as Scales himself indicates, why would we look for their presence in cities?

If, to use Miquel Barceló's epigram, 'the tribal ambiance creates tribes',[56] we must now widen our inquiry beyond conversion and inquire by what mechanisms indigenous settlements might have come to emulate those of segmentary society. Is it really logical to assume that a detribalised late Roman population would become tribalised? What does 'tribal' mean in such a context?

We first need to theorise the kinship structure of the indigenous population on the eve of their subjugation to a tribal society. Scholars do not agree on the status of late Roman kinship structure. For Goody, the *gens* had disappeared in Roman Spain.[57] But is there any real way to determine to what extent bilaterality had replaced it? Pierre van den Berghe states the problem thus:

> The question is: when did the Romanised Iberians become bilateral? Very possibly they were still patrilineal in the 10th or 11th century.[58] Even if they had become bilateral by then, conversion to Islam might well have made them shift back to patrilineality, because of the strong patrilineal bias in

54 That clientage was widespread, there can be no doubt; cf. the Romance loanword *maulado*, and my discussion in *Islamic and Christian Spain*, p. 153.

55 Peter C. Scales, *The Fall of the Caliphate of Cordoba: Berbers and Andalusis in Conflict* (Leiden, E. J. Brill, 1994), pp. 159–60.

56 Miquel Barceló, 'Vísperas de feudales. La sociedad de Sharq al-Andalus justo antes de la conquista catalana', in Felipe Maíllo Salgado, ed., *España, Al-Andalus, Sefarad: Síntesis y nuevas perspectivas* (Salamanca, Universidad, 1988), pp. 99–112, on, p. 107.

57 Jack Goody, *The Development of the Family and Marriage in Europe* (Cambridge, Cambridge University Press, 1983), p. 17.

58 I would emend this to read 'eighth or ninth'.

Islamic inheritance, marriage, descent, divorce and other aspects of family law. . . . This also raises the question of the Berbers. Most contemporary Maghrebine Berber groups are matrilineal (and monogamous) after 1,200 years of Islam. Was this not also the case with the Iberian Berbers?[59]

For Acién Almansa, Visigothic society on the eve of the conquest was proto-feudal. The conquest set in motion a twin process: on the one hand, an assimilation of Muwallad villages to tribal norms (assimilating Arab or Berber family structure); on the other, Arabs and Berbers, in imitation of Gothic structures, began to display, in certain lineages, aristocratisation, whereby certain lineages emerged as powerful and controlled other tribes and territories. This transition ends with the *fitna* of the late ninth century. At that time, Muwallads lived in fortified communities (that is, *ḥiṣn/qarya* complexes) in the mountains of Granada and Almería and Ibn Ḥafṣūn had to bring a number of these under control. (Acién Almansa adds that this process resembled that whereby the Carolingians controlled the free villages of Austrasia.) Ibn Ḥafṣūn's political relationship with castles and dependencies other than his own resembled feudal ones: that is, the Muwallad *ṣāḥib*/s of various castles appear to have had a relationship of dependency on Ibn Ḥafṣūn. The defeat of Ibn Ḥafṣūn by ᶜAbd al-Raḥmān III, therefore, represented the end of the last 'lords' of indigenous origin, who represented a particular kind of patrilineal social organisation already present in Hispano-Roman society in 711 and which he characterises as proto-feudal.[60] It is important to note, assuming that bilaterality had made substantial inroads among Hispano-Romans on the eve of conquest, that bilateral kinship and patrilineal descent (clan organisation) are not necessarily incompatible.[61] With respect to Muwallad/s, therefore, it may well be that they 'tribalised' to the point of identifying patrilineal descent groups, but without necessarily losing their bilaterality thereby. By such a device, Muwallad/s could have adjusted to a new set of dominant social norms which prized patrilineality (which in any case would have been imposed, through Islamic family law, on all Muslims).

The final phase of the Paleoandalusi period overlaps the onset of Bulliet's 'early majority' phase of conversion, when the society

59 Personal communication, 30 September 1975.

60 Manuel Acién Almansa, 'Poblamiento y fortificación en el sur de Al-Andalus. La formación de un país de *ḥuṣūn*', *III CAME*, I, 135–50, pp. 142, 144; see comment by Miguel Barceló, 'Quina arqueologia per al-Andalus?', in *Coloquio Hispano-Italiano de Arqueologia Medieval*, pp. 243–52, on p. 246.

61 Goody, *Family and Marriage*, p. 210, specifies instances of societies in which men and women inherit from one another.

was about to be swamped by masses of Christian converts. In terms of historical events, this phase is bracketed by the martyrdoms of Córdoba in the 850s and concludes with Ibn Ḥafṣūn's revolts of the late ninth and early tenth centuries. The martyrdoms have recently been analysed by Jessica Coope, who sees the primary social dynamic of the phenomenon in terms of a split in the Christian community, threatened by the early signs of mass conversion.[62] The biographies of many of the martyrs reveal religiously mixed families (a very recent phenomenon in the 850s) and a pattern whereby men converted (some for reasons of professional preferment and prestige) and women tended to lag. Nor was the phenomenon limited to Córdoba: in 851 the *walī* of Huesca ordered the execution of two Christian youths, Nunilo and Alodia, children of a father converted to Islam and a Christian mother. The *walī* himself, incidentally, was probably a member of the Banū Qasī family, Muwallad, ex-Gothic aristocrats converted in the eighth century.[63]

Conversion also involved a gradual taking-on of Muslim identity: Arabised Christians were able to 'pass' as Muslims and many were circumcised (required for government positions), obeyed Muslim dietary customs, dressed in the Muslim fashion, shared the common Muslim view of the nature of Christ's divinity, and regarded Christianity and Islam as offering no substantial differences. Such attitudes promoted a split in the Christian community between assimilationist Christians and the radicals from whom the martyrs were drawn. The pattern of martyrdoms therefore can be shown to be consistent with the dynamics suggested by the 'curve of conversion'. Moreover, the gradual nature of conversion that Coope suggests was the norm in Córdoba is a further indication that the pace of conversion was slower rather than faster. This is a much more coherent psychology of conversion than that implied by those who argue for rapid mass conversions early on. The Muslims, as is well known, welcomed conversion, but did not apply direct pressure to achieve it from *dhimmī*/s. In the absence of 'push' factors, therefore, sufficient time must be allowed for 'pull' factors to build up. Such factors were pre-eminently social and acquired real social and political meaning at the point when conversion began to split Christian families in significant numbers. The vehemence

62 Jessica A. Coope, *The Martyrs of Córdoba: Community and Family Conflict in an Age of Mass Conversion* (Lincoln, University of Nebraska Press, 1995). In a related article, 'Religious and Cultural Conversion to Islam in Ninth-Century Umayyad Cordoba', *Journal of World History*, 4 (1993), 47–68, the mechanisms of family interaction are briefly described on p. 58.

63 Carlos Laliena and Philippe Sénac, *Musulmans et Chrétiens dans le Haut Moyen Age: Aux origines de la reconquête aragonaise* (Toulouse, Minerve, 1991), pp. 37, 49.

of the radicals was directed not against Muslims but against assimilationist Christians.[64]

The best evidence (as opposed to pure conjecture), therefore, supports the dynamics of Bulliet's curve. Sufficient grounds for 'reperiodising' Bulliet have not yet been established.

Religious and 'social' conversion

Paleoandalusi culture begins with the Muslim conquest of Hispania and ends with a substantial roster of 'early majority' phenomena defining the most active (and conflictive) stage of acculturation/ assimilation. Conversion, as we have seen, occurs by a process of contagion and is associated with social–psychological phenomena related to 'push' and 'pull' factors. As more and more people convert and Islam becomes the normative religion for a majority of persons, the weight of 'pull' factors obviously becomes stronger, since among such factors are prestige, access to public positions, a more efficient way to enter the market-place, and avenues for social mobility that were closed to *dhimmī*/s. Nevertheless the pace of acculturation is slow because people, especially those in traditional societies, will change as little of their lifestyles as is consistent with the accommodation they seek to make with the dominant group.

Bulliet distinguishes between religious conversion *per se* and 'social conversion'. In this view, formal conversion to Islam meant very little unless accompanied by the attachment of each individual convert to a new religious community, a distinction which only has meaning in societies where social identity is defined in religious, rather than in tribal, terms.[65] In Al-Andalus, therefore, social conversion (in Bulliet's terms) was crucial. This explains why the Epalza–Llobregat thesis doesn't work: even if large numbers of Hispano-Romans were converted *en masse* because they were pagans, such a conversion means little in social terms, nothing in cultural terms.

What is the relationship, then, between such apparently disparate cultural elements as pots, astrological methods, language, and religion? Are they all independent variables, or should religion be privileged as the independent variable on which the others depend, and to what extent?

Bulliet's notion that early converts to Islam tended to migrate to cities helps explain differential rates of acculturation and why,

64 Coope, *Martyrs*, pp. 24–30.
65 Bulliet, *Conversion to Islam*, p. 34.

for example, forms of material culture that archaeologists have identified as late Roman should have persisted in rural areas. Muslim institutions developed in cities, and rural Islam developed substantially later, 'partly because the remoteness of the countryside slowed the pace of conversion'.[66] These early converts tended to be staunch supporters of the regime and were quickly assimilated into the nascent urban elite. This set up a situation of social conflict between early and late converters, a dynamic that perhaps can be seen in the *fitna* (rebellion) of the late ninth and early tenth centuries, led by the Muwallad Ibn Ḥafṣūn whose followers, if Acién is to be believed, were remnants of the Gothic nobility. That later converts, located in rural areas, should be associated with archaic technologies is a logical concomitant of this social dynamic, in which urban/rural differentiation was exacerbated by, and bent to the logic of, the curve of conversion. This dynamic of the migration of early converts to the cities lends substance to Riu Riu's assertion that large numbers of Mozarabs fled from the cities to the countryside as a result of the persecutions of the mid-ninth century.[67]

Acién Almansa's recent analysis of the revolt of Ibn Ḥafṣūn reveals a dynamic which can be viewed as a postscript to the Paleoandalusi period, and further reflects the playing-out of an 'early majority' social scenario.[68] The *fitna* has been perceived in the past as a movement with ethnic undertones: in this view, Muwallad/s made common cause with Mozarabs in order to protest a polity in which power was concentrated in the hands of an Arab Muslim elite. But Acién Almansa shows, first, that Muwallad social structure was more complex than had been thought and that patterns of political alliances during the *fitna* do not reveal an ethnic or religious rationale. Nor was the *fitna* a popular movement: the urban masses, of whatever ethnic background, tended to support the ruling Umayyads, while rural Mozarabs (Acién Almansa supposes) were lukewarm at best about supporting Ibn Ḥafṣūn. The Muwallad rebel leadership attacked people of all ethnic groups

66 Bulliet, *Conversion to Islam*, pp. 55–6, 81.

67 Manuel Riu Riu, 'Aportación de la arqueología al estudio de los mozárabes de Al-Andalus', in Riu Riu *et al.*, *Tres estudios de historia medieval andaluza*, 2nd ed. (Córdoba, Caja de Ahorros, 1982), pp. 82–112, on p. 91: after the mid-ninth century the Mozarabs preferred to live in easily defendable mountain settlements. See also his earlier article, 'Poblados mozárabes de Al-Andalus. Hipótesis para su estudio: El ejemplo de Busquistar', *Cuadernos de Estudios Medievales*, 2–3 (1974–75), 3–35, on pp. 6–7, where he identifies the reign of the Emir Muḥammad as decisive for the fortification of Mozarab settlements.

68 The following discussion is based on Manuel Acién Almansa, *Entre el feudalismo y el Islam. 'Umar ibn Ḥafṣūn en los historiadores, en las fuentes y en la historia* (Jaén, Universidad, 1994).

indiscriminately, either for the purpose of plundering them or to impose tributes upon them. Likewise Ibn Ḥafṣūn attacked other Muwallad rebels and made allegiances with powerful Arab families like the Banū Ḥajjāj and Banū Khaṭṭāb.

The Muwallad rebel leaders typically controlled large *ḥiṣn* complexes (*ummahāt al-ḥuṣūn*), some of which dominated agricultural districts. These leaders all belonged to cohesive family groups, in which sons succeeded fathers to control of *ḥuṣūn*. They were supported by *ṣāḥib*/s, who controlled smaller castles in a network of *ḥuṣūn* established during the *fitna*. Power circulated within the rebel group in a social hierarchy through some kind of internal promotion mechanism that is undocumented. Both rebel chiefs and secondary *ṣāḥib*/s and retainers were all descended from the Gothic aristocracy. Ibn Ḥafṣūn's great-grandfather was a man named Marcelo, who had been a notable in Visigothic Ronda. His grandfather, Jaᶜfar al-Islāmī (nicknamed, 'the convert to Islam') was said to have 'become great' (possibly a reference to the wealth and power he derived from his estates), adding the honorific suffix *-ūn* to the family name. The remnants of the Gothic aristocracy sought to fight off the weakening of lineages, to which mixed marriages were the greatest threat. At the Council of Córdoba in 839 the Mozarab hierarchy had prohibited them – in itself a vivid sign that such were occurring on a massive scale.[69] The Muwallad lineages that appear in the *fitna* were those that had maintained the male lines intact. But even elements of the Gothic aristocracy that were absorbed into the Arab patrilineal system were represented: the Banū Ḥajjāj and Banū Khaṭṭāb themselves controlled large estates by virtue of having intermarried with the Gothic nobility, the former with the family of the Visigoth King Witiza, the latter with the descendants of Theodomir, protofeudal chief of the region called, after him, Tudmir.

Therefore, the *fitna*, in Acién Almansa's view, was the final attempt of the remnants of the protofeudal Gothic aristocracy, now mainly Islamicised, to forestall the imposition of a state governed solely by Islamic norms, and which did not countenance the privatisation of jurisdictions or the substitution of rents for taxation.[70] The *fitna* failed because it lacked a social basis of support. These semi-feudal warlords had lived by exacting rent from the communities of Mozarabs who had fled the cities around the time of the

69 *Ibid.*, p. 112. I have already commented on Coope's analysis of this typical 'early majority' dynamic.

70 Acién Almansa rejects public/private distinction as valid, even while making it a key part of his argument (*ibid.*, p. 108).

Islamic conquest (not fleeing from Muslims, in Acién Almansa's view, but rather attempting to rid themselves of the oppressive protofeudal Visigothic aristocrats, from their bases in *ḥiṣn* complexes).[71] ᶜAbd al-Raḥmān III defeated the Muwallad *aṣḥāb* militarily, then separated them from their lands and integrated them into the army.

Here we have presented the contours of a Paleoandalusi period with a distinctive culture comprising, first, the material infrastructure of the garrison State, late Roman in nature and linked to specific archaic protofeudal social groups that accounted in turn for some of the political dynamics of the period. Late Roman culture in transition can also be detected in Paleoandalusi science and medicine, and in specific geographical and ideological processes associated with the peopling of a new land.

71 In a famous passage, Ibn Ḥawqal describes a mass of rural '*Rūmī*/s' (that is, Christians), who opposed the Umayyad state from rural *ḥuṣūn*, practised agriculture, and knew nothing of cities. Acién Almansa (*ibid.*, p. 122) notes that Ibn Ḥawqal's perplexity was caused by his lack of familiarity with the kind of protofeudal society he was commenting on.

4

Irrigation in Al-Andalus: a reassessment

Irrigation in the Mediterranean world

The historical study of Mediterranean irrigation systems involves a series of problems whose solution, at a conceptual level, is highly complex. Butzer and his team formulated a view of irrigation in eastern Spain (based on excavations and field surveys in the province of Castelló) which locates Andalusi irrigation squarely in a tradition of Mediterranean agriculture reaching back four millennia. They find that the great break in Mediterranean history constituted by the Islamic invasions was no break at all in so far as agriculture is concerned. Rather than sponsoring an 'Arab green revolution', the Muslims simply 're-created the Roman agrosystem, and amplified it with methods and cultivars already tested and perfected in India, Persia, Mesopotamia, Syria and Egypt'. Only nineteen of 134 'economic' plants recommended by the Andalusi agronomists were in fact introduced by the Muslims; and these supplemented a basic list of traditional staples grown by the Romans; nor can Muslim originality over the Romans be claimed in regard to crop rotations, soil fertilisation or irrigation.[1] For Butzer, schooled in the field as a geologist and prehistorian and whose major work was on the invention of agriculture, the Islamic period goes by in the wink of an eye and all we can see over the perspective of millennia is the persistence of a rather standard technological 'package'. In such a perspective cultural transitions are difficult to perceive or define, to say the least. I have argued that agrarian technology includes institutional arrangements for water distribution which both encode ecological information and provide useful cultural markers.[2]

1 Karl W. Butzer *et al.*, 'Irrigation Agrosystems in Eastern Spain: Roman or Islamic Origins?', *Annals of the Association of American Geographers*, 75 (1985), 479–509, especially pp. 500–4.

2 Butzer's footnote 6 (*ibid.*, p. 505) reflects my views on this issue, expressed in correspondence with Butzer.

On the other hand, Butzer has provided a very useful analytical tool for evaluating the Muslim contribution to Spanish irrigation development by proposing a tripartite typology of irrigation systems. Macro-scale irrigation refers to large irrigation works (from 50 to 100 sq km) found on major flood plains, such as the Ebro River or to alluvial coastal huertas, including three studied by Butzer (the Plana of Castelló, the Palancia huerta around Sagunto, and the huerta of Valencia), to which a number of others, such as Alicante and Murcia/Orihuela could be added. In these huertas, there is ample evidence of Roman irrigation, as well as of considerable extension of the irrigated surface during Islamic times. Meso-scale irrigation is that typically associated with mountain villages and hamlets watered from springs or small rivers between 15 and 125 ha. These Butzer finds to be of Muslim foundation. I might add that the mesosystem is that which describes the irrigation agriculture associated with *ḥiṣn/qarya* complexes. Micro-scale irrigation, finally, refers to small terraced parcels (around 1 ha in area) irrigating from tanks (*albercas*) or cisterns that store water from minor springs.[3] This type of hydraulic organisation is spread widely throughout the Islamic world but, Butzer concludes, 'Both the features and the method are too ancient and too commonplace to infer either classical continuity or Islamic reintroduction'.[4]

There is some evidence of continuity with Roman irrigation. In Cieza (Arabic, Siyāsa), for example, al-Zuhrī (twelfth century) states that the Romans (*Rūmī/s*) had canalised water from a spring in Minatea. This canalisation has been associated with the *acequia* de Don Gonzalo, an irrigation canal 24 km long which flows along the right bank of the Segura river, ending at the foot of the hill occupied by the *ḥiṣn* of Siyāsa. On the left bank of the river is another canal, now known as the *Acequia de Los Charcos.* Along both these canals are a number of Roman villa sites yielding ceramics from the first and second centuries AD, although it was the Don Gonzalo canal that was used by Islamic Siyāsa in the twelfth and

3 I disagree with the conclusion of Helena Kirchner and Carmen Navarro that the distinction of scale is irrelevant inasmuch as 'the principles of design of a hydraulic space are always the same'; 'Objetivos, métodos y práctica de la arqueología hidráulica', *Archeologia Medievale*, 20 (1993), 121–50, on p. 124. As I argue here, customary institutional features whose purpose is to reduce conflict-producing ambiguity in irrigation systems are also design features, and they differ depending on the scale of irrigation, e.g., the longer a canal is, the less effective will self-monitoring measures be.

4 Butzer *et al.*, 'Irrigation Agrosystems', pp. 486–99; quote on p. 499; Butzer *et al.*, 'Orígenes de la distribución intercomunitaria del agua en la Sierra de Espadán (País Valenciano)', in *Los paisajes del agua* (Valencia/Alicante, Universidad, 1989), pp. 223–8, on p. 223.

thirteenth centuries. The authors observe that wherever Romans practised intensive irrigation the infrastructure was reused; therefore archaeological study is very difficult (that is, in this case it is difficult to identify the precise course of the *acequia* of Don Gonzalo either in Roman or in Islamic times).[5] Similar evidence comes from Granada where recent excavations produced Roman remains on fields watered by the Acequia Gorda, one of the principal canals of the medieval vega of Granada. One of the fields is bisected by a secondary canal, 'a fact which suggests to us the antiquity of the Granadan irrigation system, whose dating (at least that of some of the main canals) is increasingly accepted as Roman'. The *villae* and *vici* of Roman Granada (Iliberis) were located alongside roads or irrigation canals.[6]

In the sterile debate over the 'origins' of Spanish irrigation systems, historians and publicists of the nineteenth century chose Rome as the most likely ancestor if they represented a centralist position and supported a series of water laws inspired in the French civil code. Advocates of the traditional, autonomous local irrigation communities opted for Arab origins. The former based their arguments on archaeological evidence of Roman water systems, the latter on the prevalence of Arabisms in the terminology of irrigation. But the debate was ideologically driven and generated a lot of heat and little sound evidence.

From the perspective of this book the question of 'origins' is interesting only in so far as it touches on cultural continuity or discontinuity during the transition from Roman to Islamic social structures.[7] In assessing the roots of the irrigation systems established by Muslims in Al-Andalus, the historian confronts a paradox: an extreme mixing of cultural forms in a relatively short time span (seventh to eighth centuries), followed by the hyperstability that

5 Antonio Yelo Templado *et al.*, 'Aportación al estudio del poblamiento y los regadíos de época romana en la cabecera del valle del Segura, fuentes documentales y arqueológicas', *Antigüedad y cristianismo* (Murcia), 5 (1988), 599–611. These authors identify al-Zuhrī's *madīnat Iyih* with the Iyih mentioned in the 'Pact of Tudmir'. But Tolmo de Minatea (Hellín) is a geographically more logical choice (Alfonso Carmona, personal communication, 9 December 1993; Sonia Gutiérrez Lloret, 'La formación de Tudmir desde la periferia del estado islámico', II jornadas sobre Madinat al-Zahra: Al-Andalus antes de Madinat al-Zahra (Córdoba, 1991), typescript, p. 9).

6 Manuel Espinar Moreno, Juan José Quesada Gómez and José Amezcua Pretel, 'Materiales romanos, visigodos y árabes en la autovia de circunvalación de Granada. Aportaciones a la arqueología y cultura material', in *In Memoriam J. Cabrera Moreno* (Granada, Universidad, 1992), pp. 103–16, especially p. 115.

7 For the nineteenth-century antecedents of the current historiographical debate, see Thomas F. Glick, *Irrigation and Society in Medieval Valencia* (Cambridge, MA, Harvard University Press, 1970), pp. 149–74.

irrigation systems typically display, which creates the illusion of immutability and permanence. Hyperstability presents an opportunity and a risk. The opportunity is that cultural elements of differing origins are frozen in time for us to examine now and reconstruct their past. The risk is to see more coherence than might have been the case. A culturally complex and new society like that of early Islam cannot have presented any unified set of institutions. The Arabs pulverised the societies of antiquity and adopted cultural elements torn from their original contexts. Tribal irrigation practices, whether Arab or Berber, no doubt display greater coherence than other areas of culture. However, Berber and Arab societies also had generalised rules and prescriptions which pertain to customary law and which may also have displayed the remnants of Roman provincial law of the Near Eastern and North African provinces.[8]

A local control model of irrigation in Al-Andalus

Karl Wittfogel's hypothesis of 'Oriental despotism' supposed (following an idea of Marx's) that the water control requirements facing the ancient societies of the Near and Far East generated centralising, despotic structures which he felt to be inherent in the nature of water control. Subsequent research, stimulated by his hypothesis, demonstrated that the argument has some validity, but only if the cases are limited to control of large rivers whose regimes are characterised by seasonal flooding in arid regions. Once the focus of study changes to smaller water sources or to semi-arid regions, the institutional array is bewilderingly complex.[9]

Oriental Despotism remains an important work, not so much because of its global theories of political organisation but rather because, first, it provides a ready-made framework for comparative analysis, even of places *within* the Islamic world which include irrigated regions representing different kinds of climatic situations from arid, to semi-arid, to temperate. Second, Wittfogel focuses the

8 See Patricia Crone and Michael Cook, *Hagarism: The Making of the Islamic World* (Cambridge, Cambridge University Press, 1977), p. 94 (bits and pieces of Greek thought, odds and ends of Roman law 'all torn from their original contexts to provide materials for an Arab edifice'); Patricia Crone, *Roman, Provincial and Islamic Law: The Origins of the Islamic Patronate* (Cambridge, Cambridge University Press, 1977).

9 See my discussion of Wittfogel in my 'Irrigation and Hydraulic Technology in Medieval Spain', Variorum, in press.

problem of irrigation on its social and political requirements and organisation. His hypothesis takes ecological adaptation (which necessarily involves the application of technology) both as the starting-point for the elaboration of a typology of social organisation and as a means to explain cultural and social change.[10] It legitimises in a very broad context the historical study of irrigation and hydraulic technology. Finally, in deliberately promoting a 'universalising' hypothesis, Wittfogel pushed his theory beyond the bounds of arid civilisations, which had the effect of enhancing the testability of his thesis.[11] The result has been a literature of extraordinary richness, detailing variations both in institutional forms of water control as well as the widest possible range of political contexts of irrigation.

Wittfogel, based on the insufficient sources at his disposal, wrongly concluded that Al-Andalus was a hydraulic society. In fact, it was one in which irrigation was characterised not by bureaucratic organisation and centralised control, but rather the opposite: tribal organisation and local control.

A different approach to the role of water in social and political organisation of ancient civilisations can be traced not to Marx, but to Hegel, whose construction is found in the Appendix on the geographical basis of world history in his *Lectures on the Philosophy of World History*. Here he describes how early centres of civilisation, embodied in independent states, arose in broad river valleys, where the 'gradual accumulation of silt has made the soils fertile': 'The river plains are the most fertile lands; agriculture becomes established there, and with it, the rights of communal existence are introduced. The fertile soil automatically brings about the transition to agriculture, and this in turn gives rise to understanding and foresight.'[12] Hegel did not specify irrigation nor was he aware of the institutional specificity of irrigation agriculture. His conclusion that river basin agriculture stimulates the creation of 'rights of communal existence' seemingly contradicts Marx's statement that the scale of broad river valleys was too large to generate voluntary associations. According to Hegel, alluvial agriculture provided the primary

10 See Jacinta Palerm Viqueira, 'Sistemas hidráulicos y organización social', Symposium on Hydraulic Systems, Modernisation of Agriculture and Migration, Toluca (Mexico), 1991 (typescript). Palerm points out (p. 8) that Wittfogel distinguishes between 'unicentric' complex societies, such as those of pre-classical antiquity, where irrigation generated despotic forms of government, and pluricentric societies, where a variety of political and social structures were generated.

11 Bell, 'Universalisation in Archaeology', p. 149.

12 Georg Wilhelm Friedrich Hegel, *Lectures on the Philosophy of World History* (Cambridge, Cambridge University Press, 1975), pp. 158–9.

stimulus to the growth of high civilisations and created the communal solidarities which water management requires.

Among present-day historian/archaeologists the decentralised option has been modelled in an interesting way by Miquel Barceló, according to whom there is an admirable fit between segmentary (tribal) social organisation and the nature of hydraulic space in microsystems and mesosystems. The fit is owing to the structural requirements of the 'line of rigidity' of these small systems, understood as the line described by the contour of the main irrigation canal from the part nearest its source. As elaborated by Martínez, in mountainous irrigation systems, water and land require a rigid social ordering of space; thus, 'there is a perfect fit between "hydraulic microspace" and "strong" tribal parameters: demographic unity [= membership in the same tribal segment], castral unity [= political unity, in the form of dependence on a castle], hydraulic unity. In a word: the self-sufficiency of a human social group, displaying political and economic solidarity'. Such groups are presumed to be socially egalitarian: they display some inequities, but not to the point where some members have power over others, as formalised in rent. The fit between irrigation and tribal social organisation, as defined recently by a social scientist, has to do with the social requisites of the level of co-operation that irrigation requires:

> Not surprisingly, cooperation works better in small groups with similarity of needs and clear boundaries, and shared norms and patterns of reciprocity. In such communities monitoring is easier, the 'common knowledge' assumptions of models of strategic decisions is likely to be more valid, and social sanctions are easier to implement through reputation mechanisms and multiplex relationships of face-to-face communities.[13]

Barceló's model is formulated, point by point, in contrast to his parallel construction of a corresponding feudal model which, in all basic features, is the opposite of the tribal one.[14] He illustrates his general model with a heuristic model of the relationship between irrigation and milling in feudal and tribal organisations. In the former (using Catalan feudal mills as the point of comparison),

13 Pranab Bardhan, 'Symposium on Management of Local Commons', *Journal of Economic Perspectives*, 7 (1993), 87–92, on p. 90.

14 Miquel Barceló, 'La arqueología extensiva y el estudio de la creación del espacio rural', in Barceló, ed., *Arqueología medieval: En las afueras del 'medievalismo'* (Barcelona, Crítica, 1988), p. 213; Miquel Barceló, 'El diseño de espacios irrigados en Al-Andalus: Un enunciado de principios generales', AZA, I, xiii–1, on p. xxix; Miquel Barceló, *The Design of Irrigation Systems in Al-Andalus* (in press). Luis Pablo Martínez Sanmartín, 'El estudio social de los espacios hidráulicos', *Taller d'Història* (Valencia), 1 (1993), 90–3, on p. 91.

the mill is located at the head of the water diversion system, and only after it is powered does the *subtus rego* deliver water for irrigation. In Al-Andalus, on the contrary, the mill is at the end of the system, farthest from the water source, and milling is always subsidiary to irrigation. The last canal in the system delivers water to the mill, not the first, as in feudal Catalonia.[15]

There are a number of variations on Barceló's siting scheme. Kirchner and Navarro have described a very interesting and apparently widespread variation on the *alquería* pattern in Mallorca and Liétor (Albacete), respectively. These are long canals, divided into segmented modules, with each unit terminated by a mill at the lowest part of its segment of the canal. Ostensibly, each segment was settled by a clan group. Here again we denote the mill's distinctive relationship to the irrigation system which, however, has a clear hydraulic rationale as well as being congruent with communal values: the tail-end emplacement of mills provides them with the greatest possible drop on the territory of the particular clan. Such zonation also enhances self-monitoring and may act to offset head-ender/tail-ender rivalry.[16] The second variation, which is in reality a generalisation from Barceló's original model, is that of Sergi Selma, who describes two systems where the mills are located at 'neuralgic' points in the canal system (such as the derivation of the main canal, after a divisor, or at the head of a branch canal). The mills, in all cases, are sited on the main canal which regulates the flow of water into the delivery tank (*cup*).[17] Selma's scheme is close to a more general description by Ponce Molina of the mills in Fondón (Almería), where there was only one mill on each canal. Each mill typically received all the water of the river (as Selma notes with regard to the Sierra de Espadán mills, it was characteristic of this design style that the main canal regulates the flow of water to the mill, rather than siting mills on secondary branches). Below each

15 Barceló, 'Arqueología extensiva', pp. 241–2.

16 Helena Kirchner, 'La construcció de l'espai pagès: Les valls de Bunyola, Orient, Coanegra i Alaró a Mayurqa' (unpublished doctoral dissertation, Universitat Autònoma de Barcelona, Bellaterra, 1993), p. 105 and schematic diagram in Fig. 88. There is a divisor just before the delivery tank to divert the water back into the main canal when the mill is not milling. On Liétor, Carmen Navarro, personal communication, Barcelona, September 1993.

17 Sergi Selma Castell, 'El molí hidràulic de farina i l'organització de l'espai rural andalusí: Dos exemples d'estudi arqueològic espaial a la Serra d'Espadà (Castelló)', *Mélanges de la Casa de Velázquez*, 27 (1991), 65–100, on p. 71. An earlier version of this research, 'La integración de los molinos en un sistema hidráulico: La alquería de Artana (Serra d'Espadá, Castelló)', *AZA*, II, 713–36, is too categorical into imputing 'false' priorities to the builders of post-Islamic mills; see, e.g., p. 727.

mill, half of the water was returned to the river, the other half continued to flow in the main canal.[18]

Selma supposes (as does Barceló) that irrigation systems are carefully planned, that the design of the main canal is the first priority in establishing an irrigation system, and that mills included in the original design are sited in such a way as not to conflict with irrigation. Therefore, if a mill does not generate conflict, it is presumed to have been included in the original design. If it creates conflict and its use causes some terraces to be prejudiced to the benefit of the mill, that suggests the embodiment of objectives not consonant with those who planned the canal, an intromission into the original design.[19]

Selma first describes the irrigation system of Ahín (Sierra de Espadán, Castelló), a typical mesosystem of one main canal fed by diversion from a spring. This system is an *alquería* in the zone of a *ḥiṣn*/refuge (*castrum et villa*, in a document of 1260) and had four mills, all of which were part of the original design: they can only mill when water is circulating through the irrigated zone. These mills also have a clear hydraulic rationale: they slowed the current down. Given the slope of the canal, without them, the canal would have been destroyed.[20] The second system is that of Artana, presumed to be Berber because of its name and because it is another typical *alquería*, irrigated, in the zone of a *ḥiṣn*. Three mills were studied; these are also presumed to have formed part of the original system. Downstream from the mills are three tanks which store irrigation water.[21]

The problem here, of course, is that the historian is imputing motive and objectives to water users merely on the basis of the physical structure of the system and an assumed priority of irrigation over milling or vice versa. In traditional irrigation systems, however, agricultural use of water over milling is normally enforced institutionally, by establishing priorities, rather than by location.[22] Onc might expect such priorities to be socially and culturally congruent with the groups that define them, so Barceló's hypothesis might indeed be recast on the basis of priorities and irrigator objectives with some expectation of success. With what we know of Andalusi mills, many were owned by shareholders (communal

18 Pedro Ponce Molina, *El Espacio agrario de Fondón en el siglo XVI* (Fondón, Ayuntamiento, 1984), p. 71.

19 Selma, 'Molí hidràulic', pp. 72–3.

20 *Ibid.*, pp. 75–84.

21 *Ibid.*, pp. 85–95.

22 Selma's hypothesis could be easily tested by identifying systems where conflict between millers and irrigators was litigated in the centuries after the conquest.

model), while others were owned by important officials (privatised model).

Martínez also posits that hydraulic factors influence the model: that in mountain microsystems disposing of small amounts of both water and land, the social ordering of hydraulic space will be much less flexible than in a macrosystem like the huerta of Valencia, where abundance of both water and land has made the siting of mills with respect to irrigation irrelevant.[23] In the post-conquest Christian huerta, although mills were nominally feudalised, competing water-use interests were very closely regulated by the irrigation communities and seigniorial infractions of the system were actionable by peasants in the governor's court. Conflict between millers and irrigators tended to occur mainly during turns established when water was short, causing scheduling problems. In such periods, the city, not feudal lords, decided when the mills would run.[24]

It is doubtful that siting alone can resolve a social priority in water use. Moreover, the two siting modes suggest the salience of different hydraulic parameters. In the case of a tail-end mill in Almería, where mountainous water regimes are characterised by extremes of flow and a high level of stochasticity, an obvious reason to so locate a mill is to protect it from impetuous flow and to slow the current down so that it is usable to the miller: if the water reaches him in a turbulent wall he would be unable to use it all and water would be lost downstream.

Martí bases the feudal/tribal distinction less on location and more on the disposition of the component parts of the mill and its water delivery systems. In feudal Catalan mills, the storage tank (*balsa*) and delivery tank (*cup*) are contiguous and difficult to use for irrigation. In Muslim mills, the delivery tank is separate from the storage tank and water can easily be diverted for the irrigation of terraces.[25] Christians built similar mills in late medieval Seville, with mill and dam separate, creating a space for irrigation as called for in Martí's model, no doubt for reasons other than servicing the common good.[26] Indeed, Kirchner notes,

> the emplacement of the mill is not sufficient to guarantee the preference for irrigation over milling; rather than simultaneously the distribution of

23 'Estudio social de los espacios hidráulicos', p. 91.

24 Glick, *Irrigation and Society in Medieval Valencia*, pp. 137, 139–40.

25 Ramon Martí, 'Hacía una arqueología hidráulica. La génesis del molino feudal en Cataluña', in Barceló, ed., *Arqueología medieval*, pp. 165–94, on p. 170.

26 Magdalena Valor Piechotta, 'Molinos hidráulicos de rodezno en el Aljarafe sevillano', *AZA*, II, 737–52.

water [the allocation system, that is] contemplates a preference for irrigation, bearing in mind that the simultaneous functioning of more than one mill does not permit any extraction of water from the canal for purpose of irrigating. The height of the mills' delivery tanks [*cups*] presupposes that they could only have been built *after* the canal reached the necessary level over that of the stream, the lower of the two gradients.[27]

I would now like to recast Barceló's model in a more institutionally sensitive form as a general model of local control in water management, particularly as explicated by Arthur Maass.[28] In a comparative study of three Spanish irrigation systems (Valencia, Alicante and Murcia-Orihuela) and three American ones, Maass found that local control was the dominant characteristic of all six systems, 'regardless of the nationality or religion of the farmers, the epoch, whether formal control is vested in an irrigation community or in higher levels of government, the forms of government at higher levels, and perhaps even the legal nature of water rights'.[29] Local control would therefore seem to be a universal property of irrigation systems except in an environmentally narrow range (large river systems in arid, rather than semi-arid, environments) where Wittfogel's model applies. The first corollary of the local control model is that its strength and coherence 'appears to be correlated with an irrigation community's success in limiting or stabilizing growth, thereby gaining security for its members. The principle that irrigators have used universally for this purpose is time-priority, . . . but application of the principle has varied with the operating procedures for distributing water.'[30] By limiting growth, Maass means restricting water use to those who hold primary rights.

Tribal management of irrigation would seem admirably suited for this exclusivist objective, which no doubt is fully congruent with the underlying rationale of segmentary organisation which is the mobilisation of resources in favour of an agnatic group. When that organisation is applied to irrigation – as in the case of a single tribal segment organising a single canal (whether in macrosystem like those of Gandía or Murcia, or in a typical *alquería* canal) – segmentary organisation gives legitimacy to keeping water rights within the segment. An early eleventh-century *fatwa* (judicial decision) from

27 Helena Kirchner, 'La construcció de l'espai pagès'.

28 Arthur Maass and Raymond L. Anderson, . . . *And the Desert Shall Rejoice: Conflict, Growth and Justice in Arid Environments* (Cambridge, MA, MIT Press, 1978), pp. 366–76. In this collaborative work, Maass's contribution was the comparative analysis of irrigation systems, Anderson's the computer model which tested the results.

29 *Ibid.*, p. 366.

30 *Ibid.*, p. 368.

Qairawan describes a watercourse which is the indivisible property of a clan; no individual member could claim ownership of a determinate portion of the water.[31]

Both Maass and Barceló reject a rational choice model of irrigation: that is, irrigators do *not* choose the most economically rational option but rather the option (or hierarchy of options) most congruent with socially defined objectives. Note that in irrigation systems which display a high degree of institutional stability over long periods of time, the irrigators may appear not to be exercising any choice. But choice is implicit in the model.

The essence of tribal irrigation in the mesosystems we have been describing is that the rules are self-enforcing and no administrative or enforcement structure is required other than normal tribal procedures.[32] If the administrative procedures are simple, both measurement systems and turn or rotation procedures tend to be sharply structured. The purpose of the structure is to reduce ambiguity in water allocation, so that informal monitoring informed by tribal norms of co-operation is all that is required to keep stealing water to a minimum. Thus, in a Granadan *fatwa* the five canals of *Ḥiṣn* Shīrūz were each organised in rotations from the head to the tail of the canal with fixed hours of delivery.[33] In this typical rotation situation, irrigators in direct contact monitor each other: the irrigator finishing a turn would like to extend his time, but is deterred by his successor in the rotation, who (in like manner) is deterred by his predecessor from beginning his turn early.[34]

Furthermore, equality in parcel size, pretty much standard in Berber tribal irrigation, enhances the success of self-policing. The smallness of the systems also reduces the possibility of conflict between head-enders and tail-enders, which is the most frequent mode of structurally-generated conflict in irrigation. Placing a mill at the end of system practically eliminates the need for irrigators to monitor it. Therefore, norms of tribal governance are embodied in both the design of the allocation procedures and that of the physical layout of the system. A tribal group that has practised a fixed-time rotation schedule in North Africa would, after migrating to Al-Andalus, seek

31 Vincent Lagarderè, *Campagnes et paysans d'Al-Andalus (VIIIe–XVe s.)* (Paris, Maisonneuve et Larose, 1993), p. 268: *nahr mushā*ᶜ *bayna qawm.*

32 The following considerations are broadly based on Franz Weissing and Elinor Ostrom, 'Irrigation Institutions and the Games Irrigators Play: Rule Enforcement without Guards', in Reinhard Selton, ed., *Game Equilibrium Models, II: Methods, Morals, and Markets* (Berlin, Springer-Verlag, 1991), pp. 188–262.

33 Lagardère, *Campagnes et paysans*, p. 285.

34 Elinor Ostrom, *Crafting Institutions for Self-Governing Irrigation Systems* (San Francisco, ICS, 1992), pp. 72–3.

to design an irrigation system whose physical characteristics are congruent with such an arrangement.

The tribal mode of political interaction, moreover, involving the constant testing of neighbouring segments, suggests that inter-community or regional arrangements among segments sharing the same water source would be brought into conformity with the broader pattern of regional politics. The rigid and absolute priority of upstream over downstream users in Berber water law is in part designed to prevent inter-tribal competition from interrupting water distribution arrangements.

In a macrosystem, we can presume the salience of cross-cutting solidarities: membership in the commons of a canal is not the same as membership in an extra-urban suburb or *alquería,* or indeed even in a common tribal segment. Under a strict tribal model, the two might well be coterminous, in their origins at least. Over time, however, homogeneous tribal control of a single canal might well evolve into a fiction, but one which is extremely functional to the objectives of local control, especially the limitation of priority water rights to members of a core group. A successor landholder, not a member of the founding segment, succeeds to the water rights of his antecessor; to maintain the fictive unity of a tribal segment is to his advantage, even though he may belong to another tribal or ethnic solidarity.

Particularly in the realm of conflict resolution, the tribal model would appear to possess singular social force and coherence. Here there would appear to have been substantial similarities shared by the irrigators of Al-Andalus and Christian Spain. In both societies conflict was resolved at the local level, with the resolution guaranteed by lower judicial or executive officials, the *qāḍi* in Al-Andalus, the bailiff or governor in the Crown of Aragón with appeal, in both societies, to the ultimate sanction of the king (as documents from both Nasrid Granada and the Christian kingdoms show). The latter in no way contravenes the norms of local control, but merely represents the State's interest in the promotion of a very successful economic sector, one which was regarded as crucial under both tributary and feudal norms of kingship.[35]

Theorists have described two polar models of the social organisation of water use. In the first, the State both invests capital in waterworks and administers them after they are built. In this model,

35 Barceló underestimates the significance of cross-cutting solidarities. The state does not necessarily 'degrade' peasant autonomy by its intervention in irrigation; it may enhance it, as when intervening against a local magnate. Cf. Barceló, 'Diseño de espacios irrigados', p. xxxiv.

local communities, while not necessarily eliminated from lower-level participation in water affairs, have no substantial authority in them. In the second model, control of irrigation is vested in local communities who retain substantial power in organising their own affairs under a variety of governmental types, no matter who may be the formal owner of water rights or land. In both Al-Andalus and Christian Spain, irrigation was organised under a local control model. In both societies, also, there was some intrusion of the centralised model as when, under unspecified circumstances, Taifa kings appointed irrigation officers, or, more significantly, when Christian kings and lords removed water-powered grist mills from communal control, effectively privatising them.[36]

Meso-scale irrigation systems

The mesosystems that Butzer studied in the Sierra de Espadán are quite small – villages irrigated by one or two springs and with a dense packing of hydraulic techniques closely associated with Muslim settlement, including terraced fields, cisterns, *shādūfs*, norias and measurement by clepsydra.[37] Thus in Ahín water from a perennial spring is stored in three cisterns and then distributed to fields on each bank in a weekly rotation. At Chóvar, one irrigation system consisted of a spring, two cisterns and canals irrigating some 9 ha; a second, distinct system irrigates around 5 ha, with spring water stored by storage dam and distributed in time-units measured with a clepsydra. This second system required that water be lifted on to some fields, and ruins of both a *shādūf* and a noria remain. From this evidence, Butzer's group concludes that 'the lift and storage technology of the sierra in particular and the Valencian irrigation sphere in general indicates a combination of classical and Islamic roots. Introduction of the animal-driven water-wheel implies greater efficiency and will have facilitated Islamic intensification of agriculture in previously unirrigated areas.'[38]

A larger mesosystem, in this case of distinctly Arab introduction, is that of Banyalbufar, Mallorca. This area of 60 ha is irrigated by water from a *qanat* and then distributed into cisterns or holding

36 There are examples of fully hybrid systems, as in British India, where the Government built and controlled irrigation works down to the local level, where communities were then supposed to be empowered. See Thomas F. Glick, 'Sir Clements Markham i l'interés britànic en el regadiu hispànic a mitjan segle XIX', in C. R. Markham, *Informe sobre el regadiu de l'Espanya de l'Est* (*1867*) (Valencia, Edicions Alfons el Magnànim, 1991), pp. 7–44, on pp. 18–19.

37 Thomas F. Glick, 'Medieval Irrigation Clocks', *Technology and Culture*, 10 (1969), 424–8.

38 Butzer *et al.*, 'Irrigation Agrosystems', pp. 491–6.

tanks of two varieties, uncovered (*ṣahrīj* in Arabic, *safareig* in Catalan) and covered (*jubb* in Arabic, *aljub* in Catalan), from which terraced fields are watered in weekly turns. The fields are embanked in terraces called *marjades*, a *marga* being the sustaining wall. These terms are thought to be derived from Arabic *ma'jil* and to have been introduced, along with the associated repertory of hydraulic techniques from South Arabia under the Banū Ghāniya in the late twelfth century.[39] The turn procedure in Banyalbufar is broadly consonant with those of South Arabia (seven-day turns, with days counted from sunrise to sunset), although it is also somewhat anomalous. In Yemen water is usually distributed to irrigators by time-units measured with clepsydras or other devices and overseen by an official. In Banyalbufar there is no formal irrigation community and no official designated to oversee the turn, which is executed by agreement of the irrigators.[40] This is a rare procedure in eastern Spain, but not unheard of.

Nor was Banyalbufar the only *ma'jil* in Al-Andalus. Another was found in Liétor (Albacete), where water was accumulated in regulation and distribution tanks and then delivered through an intricate system of canals arranged in distinct sectors, each ostensibly corresponding to a tribal segment or clan.[41] Likewise the etymology of Sierra Mágina (< Ar. Magina) in Jáen is *ma'jin*, an alternate form (according to Dozy) of *ma'jil*. There is sixteenth-century documentation showing that this district had been irrigated in the Middle Ages, the canals having been abandoned because they were located in an unstable frontier district.[42]

The *qanat*, or filtration gallery, is a Persian technique widely diffused in Islamic Spain in meso-level irrigation systems. *Qanats* are sometimes described as vertical wells, but a true filtration gallery is built to skim along a water table and is provided with vertical shafts to supply access and air to workers engaged in maintaining the tunnel. In fact, the Arabs used a variety of related techniques to transport water underground, particularly in difficult terrains, not all of which captured water by filtration, and also built filtration galleries in river beds, not a topography characteristic of *qanats*.

39 María Antonia Carbonero Gamundi, 'Terrasses per al cultiu irrigat i distribució social de l'aigua a Banyalbufar (Mallorca)', *Documents d'Anàlisi Geogràfica*, 4 (1983), 32–68; Jacqueline Pirenne, *La maîtresse de l'eau en Arabie du Sud antique* (Paris, Institut de France, 1977), pp. 21–34.

40 Carbonero Gamundi, 'Terrasses', p. 60.

41 Carmen Navarro, 'El ma'gil de Liétor: un sistema de terrazas de origin Andalusí en activo', I Congreso de Arqueología Peninsular, Porto, 1993 (in press).

42 Jiménez Sánchez and Quesada Quesada, 'Toponimia y poblamiento de los montes granadino-giennenses', p. 74. R. Dozy, *Supplement aux Dictionnaires Arabes*, 2nd ed., 2 vols. (Leiden, E. J. Brill, 1927), I, 11.

Archaeologists have recently studied *qanats* in Mallorca and Crevillent (Valencia);[43] river-bed galleries or *cimbras* in Andarax (Almería), shorter than *qanats* usually are, without breathing wells, and which are not tunnels but rather covered trenches.[44] In Cocentaina there are *alcavons*, tunnels with breathing wells, but which carry water from the Alcoi river to irrigation canals without collecting water by filtration.[45] What is striking about the recent phase of hydraulic archaeology is that the *qanat*, thought an exotic technique when Oliver Asín described the galleries of Madrid in 1957, is now seen to have been ubiquitous. According to Cressier, *qanats* are generalised in the peninsula to the east of a line running from Tudela to Huelva. Some have been discovered recently in the Alpujarras and Baeza.[46] Many of these *qanats* are very small and were built by peasant technology rather than requiring the intervention of a trained engineer. In Mallorca, *qanats* ranged from 16 m to 299 m in length; the former has but one breathing shaft. The mean height of these galleries varies between 0.5 and 2.85 m, the width from 0.5 to 0.9 m. Clearly these are not the monumental galleries of Madrid or Kirman. Their homely nature makes intelligible the probability that Spaniards introduced the technique to the New World, particularly in the case of the small *puquios* of Peru.[47]

43 See the multi-authored volume, Miquel Barceló *et al.*, *Les aigües cercades (Els qanat(s) de l'illa de Mallorca*, Palma de Mallorca, Institut d'Estudis Baleàrics, 1986; and Miquel Barceló *et al.*, 'Arqueología: La Font Antiga de Crevillent: Ensayo de descripción arqueológica', *Areas. Revista de Ciencias Sociales* (Murcia), 9 (1988), 217–31.

44 Maryelle Bertrand and Patrice Cressier, 'Irrigation et aménagement du terroir dans la vallée de l'Andarax (Almería): Les réseaux anciens de Ragol', *Mélanges de la Casa de Velázquez*, 21 (1985), 115–35, on pp. 122–3. On *cimbras* supplying the fountains of Almería from the Andarax river, see Manuel Gómez Cruz, 'Las ordenanzas de riego de Almería año 1755', *AZA* II, 1101–26. The *cimbra* feeding the Fuente Redonda had a vaulted ceiling, recognised in the eighteenth century as of 'Arabic construction' (*de fábrica árabe*); *ibid.*, p. 1106.

45 Personal inspection.

46 Jaime Oliver Asín, *Historia del nombre 'Madrid'* (Madrid, CSIC, 1958). Patrice Cressier, 'Archéologie des structures hydrauliques en al-Andalus', *AZA*, I, li–xcii, on p. lxvi. Antonio Malpica Cuello, 'Un modelo de ocupación humana del territorio de la Alpujarra: Las ta'a/s de Sahil y Suhayl a fines de la Edad Media', in *Sierra Nevada y su entorno* (Granada, Casa de Velázquez/Universidad, 1988), pp. 293–315, on p. 306 (*qanat* at Albuñol; administrative district called Pago de Canit – 'a clear expression that there had to have been a qanat there'). Juana M. Rodríguez López and Lorenzo Cara Barrionuevo, 'Aproximación al conocimiento de la historia agrícola de la Alpujarra oriental (Almería). Epocas antigua y medieval', *AZA*, I, 441–66, on p. 452; María Josefa Parejo Delgado, 'El abastecimiento urbano de Baeza y Ubeda en la baja edad media', *AZA*, II, 813–36, on p. 815 (galleries called Arca del Agua and Arca del Moro).

47 Barceló *et al.*, *Aigües cercades, passim.* Monica Barnes and David Fleming, 'Filtration-Gallery Irrigation in the Spanish New World', *Latin American Antiquity*, 2 (1991), 48–68.

Mesosystems in Valencia and Murcia are in general associated with a family of water distribution and water rights customs associated with southern Arabia and the Saharan oases. Water is typically not adscribed to land and the usufruct (and sometimes the right itself) may be alienated. Distribution is generally by time units, measured by clepsydra, sand clock or the length of shadows, before the introduction of the mechanical clock in the fifteenth century.[48] In Granada, distribution systems were related but somewhat different. There the irrigation turn (*dula*) involved a set rotation and, in the case of water owned by religious trusts (*waqf/s*), there was a charge for the water. But the water was adscribed to the parcels it irrigated.[49]

It has been suggested that if many *alquerías* were settled by Berbers, then Berber, not Arab institutions should be identified as their antecedents.[50] This is easier said than done, but at least some of the parameters of the problem can be addressed. First, as Miquel Barceló has observed, Roman irrigation in North Africa frequently involved Berber cultivators irrigating according to presumed tribal norms. On this basis, Barceló offers an ingenious interpretation of a famous inscription of an irrigation turn at Lamasba (now Aïn Merwana, Algeria), whose meaning classicists have been unable to unravel. This third-century decree describes a typical mesosystem of terraces irrigated from a spring. The parcels are arranged in sectors cultivated by tribal groups who received water in sequence, with the hour and day of the week specified. The turns are designed as being of 'ascending' and 'descending' water. Barceló noticed that irrigators in Crevillent distinguished between rising and falling water (*agua que sube, agua que baja*), terms which refer simply to the order of the turn: rising water is a turn that begins at the tail of a canal, falling, at the head. This kind of alternation is standard in drought regimes in Murcia and Orihuela and, indeed, the

48 Glick, *Irrigation and Society in Medieval Valencia*, pp. 213–15; and 'Medieval Irrigation Clocks'.

49 Manual Espinar Moreno, Thomas F. Glick, and Juan Martínez Ruiz, 'El término árabe *dawla* "turno de riego", en una alquería de las tahas de Berja y Dalias: Ambroz (Almería)', *AZA*, I, 121–41.

50 References to my inadvertence to Berber institutions: Pierre Guichard, *Al-Andalus: Estructura antropológica de una sociedad islámica en Occidente* (Barcelona, Barral, 1976), pp. 304–5; André Bazzana and Pierre Guichard, 'Irrigation et société dans l'Espagne orientale au Moyen Age', in J. Metral and P. Sanlaville, eds., *L'homme et l'eau en Mediterranée et Proche Orient* (Lyon, Maison de l'Orient, 1981), pp. 115–40, on p. 130; Miquel Barceló, 'La qüestió de l'hidraulisme andalusí', in Barceló *et al.*, *Les aigües cercades*, p. 30, note 72; Eduardo Manzano, 'El regadío en al-Andalus: Problemas en torno a su estudio', *En la España medieval* (Madrid), 5 (1986), pp. 617–32, on pp. 628–9.

Lamasba inscription may well refer to a turn instituted only in case of drought.[51]

Berbers obviously irrigated in Spain; this may be inferred from the documented continuity of irrigation systems in areas of known Berber settlement. The problem is how to establish valid criteria for analysing such systems. Archaeological remains tell us little, because systems displaying relatively simple technology, such as those of small mountainous settlements, have similar physical attributes the world over. We must look elsewhere for cultural markers. In the irrigation systems of the huertas of Gandía and Murcia there are numerous secondary canals with Beni- names. In the case of Gandía, a pattern of Beni- names is clearest on the Pellerias branch of the Vernisa river, where secondary feeders have the Beni- names of their *alquerías* (Beniopa [ᶜUqba], Benicanena [Kināna], Benipeixcar, and Benisuay [Shu'ayb]), while the tertiary feeders have romance names. Many more Beni- names are recorded in the 'Distribution of Water of 1244'.[52] Guichard sees two 'toponymic strata' here: in the centre of the huerta a nucleus of military and commercial names, with an older group of Beni- names outlying them.[53] The Beni- pattern is indicative of tribal settlement, although the names just cited are Arab. A similar pattern was recognised in the 1880s by Pedro Díaz Cassou as being indicative of the tribal organisation of irrigation in the huerta of Murcia.[54] In Jacques Berque's study of

51 On Lamasba, see Brent D. Shaw, 'Lamasba: An Ancient Irrigation Community', *Antiquités Africaines*, 18 (1982), 61–103. Shaw's solution to the *acqua ascendens* problem is excessively abstract; see p. 80, where he proposes that the terms *ascendens* and *descendens* were two arbitrary types of irrigation water, 'so divided for the purposes of calculating exact proprietary claims on the water itself'. Barceló's solution is found in his 'La qüestió de l'hidraulisme andalusí', p. 16. On alternating turns in eastern Spain, see Maass and Anderson, *And the Desert Shall Rejoice*, p. 77. But note Elizabeth Fentress's critique of Shaw for the latter's stress on 'the unchanging nature of indigenous society', a thesis which (she says) 'appears in a more vulgar form as the "*permanence berbère*" of French historical tradition'. In this view, 'The changes brought about by the Roman occupation can thus be seen to lie at the level of superstructure, leaving the indigenous populations and the productive base of the economy largely untouched'; 'Forever Berber?', *Opus* (Rome), 2 (1983), 161–75, on p. 161. The kind of moral judgement that Fentress finds in Shaw (Roman imperialism was a 'bad thing') (*ibid.*, p. 170) pervades the Spanish literature that we are reviewing here. Barceló's insistent rejection of any continuity with Roman irrigation is analogous to Shaw's stand. With regard to Lamasba, the strategy of carving a rotation schedule in stone is the ultimate form of this approach to ambiguity-reduction!

52 [Roque Chabás], *Distribución de las aguas en 1244 y donaciones del término de Gandía por D. Jaime I* (Valencia, Francisco Vives Mora, 1898).

53 Pierre Guichard, *Nuestra Historia*, II, 'La Valencia musulmana: Los siglos oscuros' (Valencia, Mas Ivars, 1980), p. 217.

54 Pedro Díaz Cassou, *Ordenanzas y costumbres de la huerta de Murcia* (Madrid, Nogués, 1889), pp. 17–18, 54–5. Julio Caro Baroja repeats Díaz Cassou's evidence and

Berber irrigation in the High Atlas, one of the basic forms of distribution is by clan unit (*gens*), and in the Rif, main canals serve villages or hamlets, while secondary ones (always with the time value of twenty-four hours in an irrigation turn) serve minor lineage segments.[55]

Tribal organisation was a readily available model for organising an irrigation canal, whether by Berbers or by Arabs. If we cannot ascribe the tribal model solely to Berbers, is there some dynamic of water distribution which is peculiarly Berber? The most salient facet of Berber irrigation is the absolute priority of upstream irrigators over those downstream.[56] Time priority is not known. The canal whose intake is located furthest upstream has the right to divert the entire debit of the river at that point. This rule probably relates, in its origins, to scarcity of water and to a regime of intermittent streams, on the one hand, and, on the other, to primitive extraction technology in clay canals in which more than half the water is lost through filtration. This rule is so absolute that it does not even admit the principle of Islamic water law according to which the upstream irrigator must return unused water to the stream for the benefit of those further downstream (a clear reflection of classical Roman riparian right).[57] Such a rule makes sense in the high mountains where there is a superabundance of water and land is scarce.[58] It does not make much sense in the lowlands.

The upstream priority rule produces another distinctive effect: at the moment a river passes from mountainous terrain to a lowland plain it is held to be a different river completely and is organised as a distinct unit, separate from the upstream portion. There are

---

also that of Pedro Ibarra y Ruiz on Elche where many *dulas* (rotations) bear Beni-names; 'Regadíos y agnaciones', *II Jornadas de Cultura Islámica* (Madrid, Instituto Occidental de Cultural Islámica, 1990), pp. 161–4.

55 Jacques Berque, *Structures sociales du Haut-Atlas* (Paris, Presses Universitaires de France, 1955), pp. 153ff.; David Montgomery Hart, 'Emilio Blanco Izaga and the Berbers of the Central Rif', *Tamuda*, 6 (1958), 213.

56 In a Granadan *fatwa* reported by Lagardère (*Campagnes et paysans*, p. 285), the *qadi* held that in the case of a conflict between irrigators, and in the absence of established rights, upstream irrigators have priority over downstreamers. In the case of two irrigation canals on the same river, the upstream canal has priority over the downstream.

57 Office du Haouz, 'Problèmes de la repartition des eaux entre les usagers' (typescript, 1962, chapter 1) [p. 23] (the text is not numbered). There is no author attributed to this document, but one can presume the report was directed by Paul Pascon, a disciple of Berque. See Pascon, 'Théorie générale de la distribution des eaux et l'occupation des terres dans le Haouz de Marrakech', *Revue de Géographie du Maroc*, 18 (1970), 3–11.

58 Hart, 'Central Rif', p. 211.

numerous examples of Spanish rivers, in areas of probable Berber settlement, that change names.[59] The case for an original Berber organisation of these rivers would be strengthened if the name changes could be shown to occur at a topographical break, as between mountains and plains. This dynamic may be reflected in the organisation of the Vinalopó river, where intercommunity disputes (in Christian times) never occurred between communities of the upper and lower sectors of the river, respectively, only between communities within each sector. The two sectors were 'totally independent'.[60]

Norias

There were two kinds of hydraulic wheel. The first was the current wheel, which lifted water from a river into an irrigation canal, moved by the force of the current alone. It required no gearing.

The second was the animal-powered noria, smaller, worked by a donkey or mule moving a horizontal wheel whose energy was rendered vertical upon engaging the potgarland wheel with its toothed rim. The water raised by the noria either flowed directly into canals irrigating a field, or was stored in a tank until the farmer needed it. The noria made it possible for a single family to produce a surplus for the market and in areas where it was common, such as La Mancha or Castelló, it must have contributed to substantial economic growth on the regional level. One proof of the noria revolution is that in an area of La Mancha in the province of Toledo, archaeological surveys reveal that an original Roman irrigation system based on dams and canals was abandoned and replaced by another system based on wells and norias.[61]

Tanks or small reservoirs (called *albercas*) serve to regulate the flow of water from the noria to the field, whether because the farmer cannot continually attend to irrigation or to providing the animal with rest or simply to permit irrigation at appropriate hours of the

59 On rivers that change name, see Pedro Ponce Molina, 'Moriscos y repobladores. El paisaje agrario de Adra en la segunda mitad del siglo XVI', in *Almería entre dos culturas* (Almería, 1990), II, 839–59, on p. 840 (Rio Grande becomes Rio Adra); Carmen Trillo San José, 'La ta'a de Andarax después de la conquista', *ibid.*, I, 413–27, on p. 427 (Río Paterna becomes Río Alcolea; Río Bayarcal becomes Río Lucainena).

60 Tomás V. Pérez Medina, 'Usos del agua y conflictividad social. El ámbito del Río Vinalopó y su cuenca en la época moderna' (typescript, November 1992), p. 112.

61 Almudena Orejas Saco del Valle and F. Javier Sánchez Palencia, 'Obras hidráulicas romanas en la provincia de Toledo', *AZA*, I, 43–67, on p. 59.

day. In the summer heat, farmers liked to irrigate at sunrise or sunset.[62]

Tanks were generally made of earth, triangular in shape, with an approximate surface of 6 × 5 × 5 m, although they are also found in rectangular, trapezoidal or circular shapes. They were oriented towards the fields, in a place intermediate between the noria track (an elevated circle where the mule was hitched) and field to be watered. Its walls were of pressed earth and were planted in canebrakes to further compact the soil and to take advantage of its humidity: cane was highly valued for roofing and other artisanal uses.

The tank received water from the canal of the noria and was supplied with three outlets: one at the lowest level to irrigate low-lying fields and to empty the reservoir; another higher up, to irrigate higher sections of the field, and another, an overflow valve, to keep the water from spilling over the sides of the tank. The first two were typically made of discarded millstones, whose central hole was used to regulate the flow of water. In the inner part of each drain was placed a cord with a rag ball attached to it. When a whirlpool formed as the tank drained this cord slipped down towards the outlet, closing it. The introduction of electrified pumps and the availability of concrete spelled the end of earthen cisterns.

In the *Repartimientos* the noria is typically recorded together with an *alberca.* In Almería in the 1490s, for example, the one-noria farm plot is amply documented, e.g., 'a house and tank and noria' (*su casa e alverca e anoria*) in one lot, or, in another, 'one *tahulla* of irrigated garden . . . which has its house and noria and tank' (*una tahulla de huerta . . . e tiene su torre e alberca e acena*).[63] Some of the tanks, however, clearly served more than just one family: 'a noria and tank, which is for all who can irrigate from it' (*acena e alberca, que es de todos los que con ella pudieren regar*).[64] It is interesting to note that in the *Repartimiento de Almería,* fig trees are highly correlated with norias: a noria and tank and three fig trees, a terrace in which there are two norias, a tank, and a fig tree, a fig tree with a noria in Biator, and so forth.[65]

62 The following description from Francisco García Martín and Thomas F. Glick, 'Norias of La Mancha: An Historical Reconnaissance in Villacañas, Spain' (in press).

63 Cristina Segura Graiño, *El Libro del Repartimiento de Almería* (Madrid, Universidad Complutense, 1982), pp. 382, 455.

64 *Ibid.*, p. 129.

65 *Ibid.*, pp. 191, 358, 486.

The association of irrigation and *Ḥiṣn/Qarya* complexes

Ḥiṣn/qarya complexes were typically associated with meso-scale irrigation systems. Al-Rāzī, in his description of Lleida, makes clear the association between *ḥuṣūn* and irrigation: the castles mentioned protect irrigated vegas.[66] In zones of irrigated agriculture, *alquerías* were always sited at the edge of the irrigated area, at the point of contact with unirrigated fields.[67] Archaeological evidence from the *ḥiṣn* of El Castellar, the Andalusi antecedent of Alcoi, a town of Christian foundation, shows that the first agrarian settlements appeared here somewhat after the middle of the tenth century. The castral area, which included six *alquerías,* was that irrigated by the Acequia de Barxell, a canal flowing directly under El Castellar, an example of the linkage between a fortified hilltop and irrigated space around AD 1000.[68] Farther north, the *ḥiṣn* of Shūn (now Vall d' Uxó) controlled seven *alquerías* irrigating from the Belcaire river and, on the basis of their names, identifiably Berber: Ceneja (Ṣanḥāja), Zeneta (Zanāta), Benigafull, La Alcudia, Beniçaat, Benigasló, and Orleyl (also called Harathurle, *ḥarāt-* meaning 'quarter'), in thirteenth-century documentation.[69] Such systems, with multiple *alquerías* drawing water from the same watercourse, require political organisation in the form of negotiated pacts among the different communities.[70] On larger watercourses, different *ḥiṣn*/s and their *alquerías* may well have been drawn into regional conflict which would have been accommodated within the politics of tribal rivalry.

Many *alquerías* had irrigation systems which were technically very simple: Bazzana and Guichard mention that of Rugat (in the castral zone of Albaida), whose agrarian system was mainly dry-farmed terraces but which had a small irrigated zone. These authors suppose that such a small irrigated area cannot be counted among 'true networks organised by a political or administrative authority' but which responds rather to the independent creation of tribal society, 'communities of free and independent peasants'.

66 Philippe Sénac, 'Notes sur les *ḥuṣūn* de Lérida', Mélanges de la Casa de Velazquez, 24 (1988), 53–69, on pp. 54–5: *muy buenos rregantios* (*ḥiṣn* of Fraga) . . . *muy buenas vegas rregantias* (*ḥiṣn* of Alcolea).

67 Josep Torró, *Poblament i espai rural,* p. 60.

68 Torró, *Alcoi. La formación d'un espai feudal (de 1245 a 1305)* (Valencia, Diputació Provincial, 1992), p. 50.

69 Bazzana and Guichard, 'Irrigation et société dans l'Espagne orientale', p. 127.

70 A point made by Barceló, by way of distinguishing between isolated irrigated *alquerías* (as in Mallorca) and *systems* of *alquerías*; 'Quina arqueologia per Al-Andalus?', in *Coloquio Hispano-Italiano de Arqueología Medieval* (Granada, Patronato de la Alhambra, 1992), p. 248.

A similar construction is applied by Torró and Segura to the irrigation system of the *ḥiṣn* of Perputxent.[71] Both these systems are technically primitive; the canal of Perputxent has no permanent diversion dam, for example (in common with numerous North African mountain mesosystems, I might add). *All* of these irrigation systems, large and small, however, were governed according to the norms of segmentary society. In the cases of Rugat (on the Albaida river) and Perputxent, on the Alcoi River, we can presume that these *alquerías* were, of necessity, implicated in the regional politics of the watercourse, where *all* users were bound in a highly interdependent system of political arrangements. For the Riu d'Alcoi, there is evidence from the Christian Middle Ages that an illegal diversion of water from the Font de Molinar in Alcoi attracted a hostile reaction from *all* downstream users, including Cocentaina, Perputxent, and Gandía, all the sites of *ḥuṣūn* in Islamic times.[72] Irrigation systems whose 'hardware' is relatively simple were not only designed but had very complex internal distribution arrangements and measurement systems. They were also parties to complex and conflictive regional arrangements, no matter how seemingly insignificant the systems in question were. Such regional arrangements would have been politically sensitive, but would also have been ruled by custom, e.g., in Vall d'Uxó, where (we can assume) all seven *alquerías* were Berber, the Berber rule of absolute priority of upstream over downstream users would have been in effect.

There is interesting evidence concerning irrigation arrangements among *alquerías* in a single castral district over partition of water from a shared source. The evidence is from late medieval Granada, but in both cases, the material comes from Spanish translations of earlier Arab documents which in turn record older arrangements. The first case concerns arrangements among the *alquerías* of the *ḥiṣn* Shant Aflīdj (castle of St Felix), one of the oldest *ḥuṣūn* documented in Al-Andalus, since Al-ᶜUdhrī identifies it as the castle of an ancestor of his who rose against the Emir Hishām I around AD 788.[73] In a document of 1304, notables of five *alquerías* of Shant Aflīdj (Picena, Beni Ozmen, Armalata, Unqueyar, and Ystarán) give witness that a sixth *alquería*, Laroles, also 'of the

71 *Ibid.*, p. 132. Josep Torró i Abad and José M. Segura i Martí, 'Irrigación y asentamientos en la Vall de Perputxent', in Míkel de Epalza, ed., *Agua y poblamiento musulmán*, pp. 67–92.

72 Thomas F. Glick, 'Berbers in Valencia: The Case of Irrigation', in *Medieval Spain and the Western Mediterranean: Essays in Honor of Robert I. Burns. S.J.* (Leiden, E. J. Brill, 1995).

73 André Bazzana, Patrice Cressier, and Pierre Guichard, *Les châteaux ruraux d'Al-Andalus* (Madrid, Casa de Velázquez, 1988), p. 45.

above-mentioned castle' had no right to the water of the Río de la Ragua, and never had rights from ancient times. In subsequent documents, various *qāḍi/s* reconfirm the allocation and, finally, in 1340 King Yūsuf I ordered that neither the allocation nor the course of the irrigation canals be altered, and subsequent kings confirmed the previous arrangements in 1375 and 1437.[74] Laroles was the *alquería* furthest upstream and may well have been claiming a right according to Berber customary law which holds that upstream users have absolute priority over those downstream; but Laroles no doubt also irrigated from the Río Laroles. The documents do not reveal the basis of Laroles's claim. It is also interesting to note that a succession of Nasrid kings did not hesitate to interest themselves in local water arrangements, providing an enlightening insight into the political dynamics of the hydraulic politics of *alquerías.*

The second case concerns allocation of water of the Abrucena river among two *alquerías,* Abla and Abrucena in the old castral district of the *ḥiṣn* of Fiñana, another castle that had been loyal to Ibn Ḥafṣūn.[75] The water in dispute was purchased by the two *alquerías* from the Sultan of Granada in 1267 or 1273. In 1356, an accord was signed by the notables of both *alquerías* awarding one-third of the water to Abla, the downstream *alquería.* The document was signed by notables and officials (*alguazil, alfaquí, el muçallan*); none of the signatories had tribal *nisba*/s. In 1386 the agreement was confirmed by the *qāḍi* of Guadix, in 1409 by Yūsuf III, and in 1420 by Muḥammad IX. Later documents indicate that the water was distributed by days and hours in a fixed turn to individual irrigators, some of whom claimed rights from around 1328, and that one day a week the water was directed to the cistern of the *ḥiṣn.*[76] Here we see a complex political dynamic. Two *alquerías* purchase what was, for all practical purposes, public water for a common irrigation canal in which individual irrigators *owned* rights to stipulated time-units of water. If this was common practice then clearly it affects our view of 'tribal democracy'. There is a marked tendency in the

74 Manuel Espinar Moreno and María Dolores Quesada Gómez, 'El regadío en el distrito del castillo de Sant Aflay. Repartimiento del Río de la Regua (1304–1524)', *Estudios de Historia y Arqueología Medievales* (Cádiz), 5–6 (1985–86), 127–57, documents on pp. 150–5.

75 On this *ḥiṣn,* see Bazzana, Cressier and Guichard, *Châteaux ruraux,* pp. 51, 59.

76 Manuel Espinar Moreno, 'Reparto de las aguas del Río Abrucena (1273?–1420)', in *Revista del Centro de Estudios Históricos de Granada y su Reino,* 2nd epoch, 1 (1987), 69–94, documents on pp. 91–4; Manuel Espinar Moreno, 'Estudio sobre propiedad particular de las aguas de la acequia de Jarales (1267–1528). Problemas de abastecimiento urbano y regadíos de tierras entre las alquerías de Abrucena y Abla', *AZA,* I, 247–66.

history of many irrigation systems towards patrimonialisation and privatisation of water rights, no matter how 'democratic' the stated values of the collectivity may be. Finally, we see that even at the late dates of this documentation, the public nature of the *ḥiṣn* was still in force, and irrigators were obliged to cede one day of water for the supply of the castle.

A final example is a water document from the *qarya* of Falix, in the district of the *ḥiṣn* of Marchena (Marshana), one of the earliest cited *ḥuṣūn* of the Alpujarras, dated 1226. This document lists forty-six proprietors of parcels irrigating from the Nacimiento river (representing a village population of between 150 to 200 persons), many with Arab tribal *nisba*/s (e.g., Qaysīi, Ḥamdānī, Ghassānī, al-Anṣārī, etc.). The document stipulates that each parcel had water for a determined time and that the water was tied to the land and could not be sold. As in the previous instances, the water rights structure was subsequently ratified by a *qāḍi*.[77]

Archaeologists (especially Cressier) have correctly insisted on the integral role played by irrigation in *ḥiṣn/qarya* complexes. But the social and political nature of that integration cannot be inferred simply from spatial arrangements or from general suppositions regarding norms of tribal governance. The evidence just presented regarding, first, conflict among *alquerías* drawing water from the same source and, second, institutional arrangements regarding allocation among individual users, even if from a later period, make it possible to describe political and social dynamics of interaction among those *alquerías* and between the *alquerías* and irrigation communities and state organisms (e.g., the sultan and the *qāḍi*). These arrangements are fairly standard and, in view of the ultrastability of irrigation systems, can be presumed to represent arrangements of long standing.

Macrosystems

Macrosystems include both large flood plains along the Ebro and Segura rivers, and the alluvial huertas like those of the Mijares (Castelló), Turia (Valencia), and Segura (Murcia). To consider only

77 Manuel Espinar Moreno, 'Población y agricultura de una alquería almeriense en los siglos XII y XIII', in *Almería entre culturas. Siglos XIII al XVI*, 2 vols. (Almería, Instituto de Estudios Almerienses, 1990), I, 187–207. On the *ḥiṣn* of Marchena, see Bazzana, Cressier and Guichard, *Châteaux ruraux*, pp. 129–33 *passim*. On the participation of *qāḍi*/s in the administration of water in the Nasrid kingdom, see José López Andrés and Faustino Martín-Caro Saura, 'Organización, distribución y problemas derivados de la administración del agua en Almería y su vega en los años anteriores a la conquista', *AZA*, II, 1017–32, on p. 1026.

the latter three huertas, irrigation was practised in all three in Roman times, although the scale of development cannot be ascertained with precision: because of the continued practice of intensive agriculture, archaeological study is difficult. Reconnaissance of the plain of Castelló by Bazzana and Guichard has revealed a network of old canals around Borriana, servicing the *alquerías* in the zone of this *ḥiṣn*; upstream, around Villareal, a new town founded by Christians in the thirteenth century, is a canal system built by the settlers. Further upstream still, and to the west, are Roman sites, formerly irrigated, but now used for dry-farming.[78] In Valencia, there are vestiges of Roman canals on both the north and south banks of the Turia, upstream at Paterna and westward. The Islamic network, formed by main canals predominantly with Arab or Berber names, surrounds the city itself and cannot be documented until the eleventh century.[79]

The Murcia system yields a picture similar to that of La Plana, based on toponymic evidence analysed by Pocklington. The canals with Arab names form a coherent network, while those with Latin names (only 22 per cent) are scattered around. This pattern is an inversion of the pattern in Lorca, where the canals with pre-Arab names (68 per cent) form the nucleus of the system while the Arab-named canals are on the periphery.[80] The pattern is suggestive, but Pocklington's sociocultural deductions are incredible: the Lorca system, where water is auctioned, is inconsistent with 'Islamic philosophy', while the Roman exhibited the proper capitalistic spirit.[81] The huerta of Orihuela evolved from and replaced the primitive, noria-based Paleoandalusi system described above, no earlier than the mid-tenth century. The canal network included a system of

78 Bazzana and Guichard, 'Irrigation et société', p. 138.

79 Butzer *et al.*, 'Irrigation Agrosystems', pp. 188–9.

80 Robert Pocklington, 'Acequias árabes y pre-árabes en Murcia y Lorca: Aportación toponímica a la historia del regadío', *Xé Colloqui General de la Societat d'Onomàstica. 1er d'Onomàstica Valenciana* (Valencia, Universidad, 1986), pp. 462–73, 'Toponima y sistemas de agua en Sharq al-Andalus', in Míkel de Epalza, ed., *Agua y poblamiento musulmán* (Benissa, Ajuntament, 1988), pp. 103–14, on p. 109.

81 *Ibid.*, p. 468. See Barceló's comment, 'La qüestió de l'hidraulisme andalusí', in *Les aigües cercades*, pp. 9–36, p. 14. In *ibid.*, p. 10, he comments on classical archaeologist Miquel Tarradell's assertion that the huerta of Valencia was Roman because the Roman were engineers and Arabs weren't. No wonder Barceló cringes when he hears the word *mentalité*! For a similar *obiter dictum* by an author who opposes what he perceives as a recent wave of *maurofilia*, see Faustino Rodríguez Monteoliva, 'Los molinos de harina en la Alpujarra de Granada, durante los siglos XVI al XVIII. Léxico, etnografía e historia', *AZA*, II, 681–712, on p. 687, where he asserts that Roman technology was much more advanced than that of the Arabs in everything relating to agriculture and water use.

drainage ditches (*azarbes*) which terminated in the Paleoandalusi marshes.[82]

The current group of hydraulic archaeologists, including Barceló, Cressier, and Martí, all doubt the antiquity of the major huertas which they suppose to have been of much smaller scale in Islamic times through the caliphate and perhaps beyond, and certainly in Roman times.[83] The argument against Roman huertas as the basis on which the Muslims then built macrosystems is that, in Valencia, for example, there would have had to have been a fairly dense population to maintain the system between the third and eighth centuries, while the abandonment of latifundias and demographic decline of fifth-century Valencia is quite well-documented. In spite of all, classical archaeologists have unearthed Roman canal gates which suggest a higher level of development than the medievalists want to admit.[84]

Irrigation and peasant production

For Barceló, the notion of Andalusi irrigation's continuity with Roman practice is meaningless unless something is said about the organisation of production. Canals and dams alone do not make an irrigation system. Slave labour would create a system quite different from those created by autonomous groups of peasants.[85] In a similar vein he notes elsewhere that agriculture, strictly speaking, does not exist: what exists are processes of peasant work defined tribally. Clan organisation *is* the form that peasant work took in Al-Andalus.[86] Thus I have taken pains to stress that tribal organisation was the universal model for the allocation of water in rural Al-Andalus.

82 Sonia Gutiérrez Lloret, 'El origen de la huerta de Orihuela entre los siglos VII y XI: una propuesta arqueológica sobre la explotación de las zonas húmedas del Bajo Segura', *Arbor*, 1995, p. 87.

83 Ramon Martí, 'Oriente y occidente en las tradicciones hidráulicas medievales', *AZA*, I, 419–40, on p. 434; Barceló, 'Hidraulisme andalusí', p. 10 (Butzer's Roman macrosystems never existed). Cressier, 'Archéologie des structures hydrauliques', p. lxxiii (no macrosystems in Al-Andalus).

84 José Llorca Rodríguez, 'Romanidad de los riegos de la huerta valenciana', in *Notas sobre la antigüedad de la agricultura y el regadío en tierras valencianas* (Valencia, I Congreso Nacional de Comunidades de Regantes, 1964), pp.103–14, especially the photograph of a *tablacho* on p. 108.

85 Barceló, 'Hidraulisme andalusí', p. 10.

86 Miquel Barceló, 'Vísperas de feudales. La sociedad de Sharq al-Andalus justo antes de la conquista catalana', in Felipe Maíllo Salgado, ed., *España, Al-Andalus, Sefarad: Síntesis y nuevas perspectivas* (Salamanca, Universidad, 1988), pp. 99–112, on p. 108.

Production drives technological innovation – the adoption and use of a new technique – although other factors enter into invention and inventivity. Barceló insists that historians redefine (at least for themselves) conventional categories like agriculture, technology, irrigation, Islamisation, etc., so that their social locus, in particular the peasantry, becomes visible. Arabists traditionally have assumed that innovation diffused from city to countryside. But how inappropriate a model for the diffusion of agricultural technology in the social system we have been describing, because it presumes that peasants had no say in the matter or were incapable of making technical decisions. But peasant cultivators were not powerless to confront or resist a hegemonic ideology. They applied singly or in groups to *qāḍi*/s to seek redress of grievances according to Islamic law.[87] They could, and frequently did, resist technological innovation, particularly in traditional agricultural societies, where they are notorious for it.[88] Peasant irrigators in many different societies have also been notably successful in preserving a high measure of local control, and in warding off the rich and powerful to do so, as presently we will see in the case of peasant irrigators in 'feudal' society, who were frequently successful in litigation in royal courts against usurpations of water by magnates.

Not only is town-based diffusion implausible in the *ḥiṣn/qarya* model, but we don't even have rural elites, enlightened or otherwise, to promote agricultural progress, as happened with some consistency in eighteenth-century Europe. In the past, I have privileged technology when discussing irrigation and its diffusion from east to west, but then I am a historian of science and technology: techniques are also ideas and they diffuse according to rules that apply to other categories of ideas. The new vision of irrigation in Al-Andalus is one highly conditioned by peasant social organisation and peasant aptitudes. Peasants too had ideas about water. They took a limited range of organisational principles – proportionality, equity or equality (operationalised in rotations), local control, tribal administrative norms – and crafted institutions congruent both with their own social organisation and the organisation of their work, but also with the design of the hydraulic system. To design a system, the same objectives were taken into consideration: which design

87 Lagardère gives many examples (*Campagnes et paysans*, pp. 270–86), as does María Jesús Viguera Molins, 'En torno a las fuentes jurídicas de al-Andalus', in *Actes, Congrès La Civilisation d'al-Andalus dans le temps et dans l'espace* (Mohammedia, Université Hassan II, 1992–93), pp. 71–8, on pp. 76–7.

88 James C. Scott, *Weapons of the Weak: Everyday Forms of Peasant Resistance* (New Haven, Yale University Press, 1985).

elements would be congruent with or appropriate to tribal norms? Finally, the techniques used to build the system were also a relatively limited roster of practices, including digging wells, constructing a limited (but surprisingly versatile) repertory of tunnels, from vertical to horizontal, with familiar construction materials, enough notions of surveying to control the slope of the main canal and locate terraces so that they could receive water, and how to build, with purely local materials and skills, simple devices like norias and horizontal grist mills.

5

The transition to feudalism

'Feudalism' is a construct invented by legal theorists and historians centuries after the institutions described had failed. Thus, there will always be debate among historians as to what processes and phenomena the term covers and, as Wickham intimates, historians from different cultures will have different modal views of feudalism, which represent their own culture's experience and which are not necessarily mutually exclusive. Discussions of feudalism, in whatever region of Europe, had always been made difficult by terminological and conceptual confusion and ambiguity which was systemic in the literature. The meaning of basic terms like fief, homage, vassal, and so forth, differed widely over time and space. More significantly, unless authors were very careful to define the scope of the institutions they were discussing, very different phenomena were lumped together under the rubric 'feudal'. At one end of the spectrum were institutional historians, like F. L. Ganshof, for whom feudalism had to do with certain juridical relations pertaining among the elite, relating chiefly to military service. At the other was Marc Bloch's conception of 'feudal society' which broadened the significance of certain concepts like dependence, extending them to the peasantry, and creating a unified vision of a 'feudal' society. Both paradigms were very successful, each stimulating a vast literature on medieval social and institutional history.

I view feudalism as pre-eminently a form of political organisation, that is, a particular mode of distributing power, in the form of jurisdiction. Therefore, in this chapter, I look at different models of the political reorganisation and then at the processes of landscape change (*incastellamento* and associated phenomena) that accompanied it.

I cannot agree with Acién Almansa, for whom the distinction between public and private power is an imposition of the bourgeois value-system. In my view, which is an appreciation of the Aragonese political system as established in Valencia, shaped by concerns formed by my reading of English medieval history, the notion of public power privately held is the most distinctive aspect of feudal politics, the distinction itself having nothing to do with bourgeois values and everything to do with the way that Roman jurisprudents conceptualised the division of power between 'public' and 'private' interests. The most distinctive aspect of the medieval Valencian 'feudal' political system was the way in which it encouraged cross-cutting solidarities:[1] autonomous irrigation systems coexisted with feudal lords, or constituted islands of autonomy, if not freedom, in a seigniorial sea, supported by the royal justice system. Economic determinists should take note that without such a mixed political system, the economic boom of late medieval Valencia would be inexplicable.

Models of change

In the 1970s and 1980s a revolution took place in the way historians of Spain viewed the social organisation of the medieval Christian kingdoms. Up to that time, it was generally believed that for a variety of reasons – in particular the conditions of freedom pertaining on the Christian–Muslim frontier – Spain was 'incompletely feudalised' except for Catalonia which, because, as the 'Spanish March', it had been included in the Carolingian Empire and had made a 'normal' transition to feudalism. The problem here was a muddled conception of comparative history. The juridical paradigm was based on the 'feudal heartland' (northern France, western Germany, the Low Countries, and the 'champion country' of England), and thus had an ethnocentric bias built into it. More recent generations of historians are much less willing to define any particular set of institutions or behaviours as normative; indeed, comparative study has the effect of dissolving normativity, as anthropologists have known for years.

Once Spanish historians, beginning perhaps with Barbero and Vigil's 1978 book,[2] realised that the basis for establishing the normative paradigm was invalid, its demise was not long in coming and

1 Manuel Acién Almansa, *Entre el feudalismo y el Islam 'Umar Ibn Ḥafṣūn en los historiadores, en las fuentes y en la historia* (Jaén, Universidad, 1994), p. 108. The cross-cutting solidarity is a notion that is unknown in medieval Spanish historiography, nor is there any term for it, judging by the difficulty that translators of several of my works have had in representing the concept in Castilian or Catalan. The new view of Iberian feudalism, while much closer to the truth than the old, is seriously undertheorised, and its avatars are in general unable to detect distinctive patternings or nuancing in what they perceive as a monolithic and villainous social system.

2 Abilio Barbero and Marcelo Vigil, *La formación del feudalismo en la península ibérica* (Barcelona, Crítica, 1978; 4th ed., 1986).

the old view of 'incomplete' feudalism now yielded to one in which Spain was just as 'feudal' as northern France had been. The problem here, however, as Reyna Pastor has judiciously observed, was that categories of analysis were hopelessly imprecise and thrown together without any care for definition of terms and concepts.[3] Barbero and Vigil strongly associated themselves with Bloch's societal conception of feudalism, but, without any rigorous definition of their logic, made dependence the key structure and refused to define feudalism institutionally in any way. But inasmuch as dependence is virtually universal in stratified societies, 'feudalism' loses all specificity. There then followed dozens of books and articles in which 'feudalism' was assumed, but not defined.

In a recent book on history and historians of medieval Spain, Peter Linehan assesses the overturn in the 'liberal' doctrine of feudalism up to around 1980, lamenting that the 'twin dogmas of normative feudalism and the fundamental uniformity of feudal institutions' have been adopted in Spain at the same time as they are being jettisoned elsewhere in Europe. Linehan, without saying so, seems to be a partisan of the institutional school, inasmuch as he complains that notions of vassalage and fief are now 'mere infrastructural details', and dismisses Julio Valdeón's 1980 conclusion that León/Castile of the eleventh to the thirteenth centuries was 'a typical feudal society characterised by dependent relationships at every level'.[4] I depart vigorously from Linehan's analysis, however, when he claims that 'The assertion that eleventh-century Spain was a feudal country establishes a claim to Spain's membership, now as then, of a European community. Between the Community and the *autonomías*, historians of Spain now revile the nation as an object unworthy of scholarly attention.'[5] This kind of judgement is nothing more than pop psychoanalysis of an undefined group of historians and there is no evidence to support it. Moreover, Catalan historians, perhaps the principal political beneficiaries of the *autonomías*, by no means revile the concept of nation, as should

3 Reyna Pastor, comment, in Pierre Bonnassie *et al.*, *Estructuras feudales y feudalismo en el mundo mediterráneo* (Barcelona, Crítica, 1984), p. 59.

4 Peter Linehan, *History and Historians of Medieval Spain* (Oxford, Clarendon Press, 1993), pp. 194, 197. He means 'superstructural' details. On p. 191, he notes that the 'emergence from captivity of Spain's academic Marxists has hastened the medievalists in their headlong flight to feudalism'. This observation, as well as his comment on the doctrinaire nature of some of the new feudalism scholars (p. 228), is true enough, but I also believe that the introduction by Marxists of a counter-model was salutary, in part because Marx's notion of feudalism was found inadequate to deal with the complexity of the phenomenon.

5 *Ibid.*, p. 192.

be obvious to anyone. From the perspective of 1994, looking back on the historiography of the 1980s, the picture is quite different, because the weight of new models, mainly of French provenance, began to make themselves felt. In particular, it is the point of this chapter to argue that Pierre Toubert's notion of *incastellamento* induced a response to a particularly compelling model among a sector of Spanish medievalists, those most identified with the new medieval archaeology.

A number of models of the feudal transition promoted by French medievalists are relevant to our present purpose. I will begin by considering, as a heuristic or descriptive model, that of the so-called 'Fiscalist' school, according to which Western European societies were 'tributary' until around the year 1000. A tributary society is one whose governmental system is based on taxes paid into a public fisc. Thus it can be neatly contrasted with a feudal society in which rents are paid to private lords. The change from a tax-based system to a rent-based one is a simple way of conceptualising the 'transition to feudalism'. But, as Chris Wickham has observed, the transition is not so neat and both systems coexisted, in 300 as well as in 700. It is really a matter of balance. At the point at which the balance tips towards rent and away from taxation, the transition has been completed. For Wickham that point was reached when the German states began to base their military systems on landownership. Once the fief was invented, rent replaced taxation, the economic balance shifted, and feudal relations became more important than 'ancient' ones, even though the latter still subsisted.[6] This is by no means a new view: its conclusion is similar to that of Lynn White Jr, in his famous discussion of the diffusion of stirrup and the invention of feudalism by Charles Martel (the stirrup made possible the heavily armoured knight, who required substantial financial resources to keep him in the field: Martel fulfilled this requirement by granting his knights lands whose rents would be sufficient to support their horses and armour).[7]

The transition to feudalism, therefore, has been conceptualised in ways that have concentrated on fiscal issues, to give substance to the key transition from tax to rent. Part of this discussion has required a theorisation of the transition from the late Roman villa, on which *coloni*, of varying degrees of freedom, worked the lands of a *possessor, patronus* or *dominus*, to the Germanic villa of

6 Chris Wickham, 'The Other Transition: From the Ancient World to Feudalism', *Past and Present*, 103 (May 1984), 3–36, on p. 20.

7 Lynn White, Jr, *Medieval Technology and Social Change* (Oxford, Oxford University Press, 1962), Ch. 1.

more or less free peasant communities which in time became the peasant village of the high Middle Ages. For the fiscalists, the *villae* of the late Roman Empire were no longer great domains but settlements or fiscal demarcations whose contributors were juridically free but economically dependent on a *dominus*, a private person to whom the State delegated competencies relating to the collection of taxes and public funds.[8]

The fiscalists have had to pay particular attention to terminology and have reinterpreted, or given 'fiscalist' twists, to a whole series of terms that are relevant to any interpretation of rural social organisation. Thus the *patronus* or *dominus* of the late Empire may have had a large or a small estate, with or without dependent peasants, but *all domini* served as delegates of public authority, acting as fiscal officers and tax farmers. The group of lands over which the *patronus* exercised this fiscal jurisdiction (which is a kind of eminent domain) was the *fundus, terra,* or *ager.* The successor *villae* of the high Middle Ages, in turn, had the same fiscal function as later Roman *fundi.*[9] The entire Empire was thus divided into *fundi*, some of which were the territories of a *vicus* or *locus*,[10] and therefore began to resemble early medieval *villae.*

In the German states, including Visigothic Spain, the *fundus* continued as a territory in which, as before, a *dominus* or lord had fiscal power delegated by the State. But in documents in which a *dominus* sold, bequeathed or otherwise alienated property, real estate was now sold in a *villa*, a unit which included fields, houses, mills and peasants. Peasants were also described as alienating property in a *villa.* These documents do not describe different levels of ownership; they merely designate the *villa* as the relevant fiscal unit. Lords depicted in documents as selling entire *villae* are only doing so in a fiscal sense, transferring eminent, not real domain.[11]

When fiscalists push their system of documentary interpretation to the extreme, the reality of the countryside melts away: in the tributary system described, the *servus* is a taxpayer, a person subjected to a *servitum*, not a slave and perhaps not even dependent.

8 Josep M. Salrach, 'Del estado romano a los reinos germánicos. En torno a las bases materiales del poder del estado en la Antigüedad tardia y la alta edad media', in *De la Antigüedad al Medioevo* (Madrid, Fundación Sánchez-Albornoz, 1993), pp. 95–142, on p. 102. Salrach, in this article, performs the admirable service of summarising the fiscalist argument as found chiefly in the works of Jean Durliat and Elizabeth Magnou-Nortier, pointing out Visigothic and Catalan co-ordinates of their argument, where appropriate.

9 Salrach, 'Estado romano', pp. 111, 112 n. 47.

10 *Ibid.*, p. 113.

11 *Ibid.*, pp. 135, 137.

A *colonus* is not a rentpayer, just a peasant, generally a landowner. A *censum* is not a rent, but a tax. *Villae* are not only villages but, more importantly, fiscal units. The *dominus* is not a proprietor but a private depository of a delegated public power. Yet, as Salrach notes, so pervasive a system of the delegation of public power was the germ of the system's destruction, pregnant with social, economic, and political implications. Therefore the continuity that the fiscalists see is only continuity in the rather narrow terms which they have imposed upon themselves. Still we are confronted with Magnou-Nortier's gloomy conclusion: if the documentation of the high Middle Ages relates mainly to fiscal matters, then the 'rural history' that has been written using this source material is a fantasy.[12]

The system depicted by the new fiscalists is not so different from the perception of continuity with ancient society advanced by Sánchez-Albornoz: counts received taxes described by a late Roman terminology (*censum, tributum*), formerly paid to the king. But in an area of middling Romanisation like Galicia, Isla Frez asks, why would a Roman fiscal system survive for so many centuries? The same terms are found in Anglo-Saxon England where, he notes, no one argues for continuity with Roman fiscality. This is to confuse a terminological tradition with an institutional reality. It is no more Roman than are the 'Roman' taxpayers alluded to in early eleventh-century documents as owing *vectigalia uel tributa* to the Bishop of Santiago. Isla notes that Hostegesis, the ninth-century Mozarab Bishop of Málaga, obliged the Christians to pay a *censum publicum* or *vectigalia* (in Rome, the tax paid for use of state lands) to aid the Umayyads when they applied fiscal pressure to the Christian community, 'using old words adapted to the new social and political reality of the ninth century'.[13] This episode is most evocative because it both makes the desired point about the use of an older terminology to describe a totally different situation than that prevailing in the Christian kingdoms of the same period, and also reflects the recognition by a Christian official – Hostegesis – of the tributary nature of Umayyad society.

In spite of its limitations, the fiscalist model is attractive in the context of the present volume. Inasmuch as we deal with comparative data, it is useful to have models that account for phenomena

12 *Ibid.*, p. 141.

13 Amancio Isla Frez, *La sociedad gallega en la alta edad media* (Madrid, CSIC, 1992), pp. 151–6. See also Christian Lauranson-Rosaz's interesting remarks on similarly 'vague' reminiscences of Roman law in the same period: *L'Auvergne et ses marges (Velay, Gévaudan) du VIIIe au Xe siècle. La fin du Monde Antique?* (Le Puy-en-Velay, Cahiers de la Haute-Loire, 1987), pp. 211–18.

on both sides of the cultural frontier. The theorists of Andalusi society also presume a tributary model, which they counterpose to the feudal model prevailing in Christian Spain after AD 1000.

The fiscalist model also dovetails chronologically with Guy Bois's account of the 'transition to feudalism' to which I now turn. According to Bois, direct fiscality, exercised by the State, had collapsed in the sixth century, but the Church then constituted itself a kind of paragovernmental fiscal system. It collected the tithe, the functional equivalent of the ancient land tax, and through the tithe a free peasantry participated in the maintenance of the upper ranks of society. Thus took place a social redistribution of wealth which had the result of slowly weakening the free peasantry.[14] The peasantry, nevertheless, was still free, and a class of rural slaves persisted. Bois opposes the view that the Roman *colonus* of as early as the third century was a serf. These rural slaves, who constituted up to 15 per cent of the population (in the area of France studied by Bois), could be freely alienated by their owners and had no rights over the land they cultivated. They were economically and socially significant, mainly in their support of the lower aristocracy.[15]

Nor was slavery the only element of 'ancient' society that survived into the tenth century. Old administrative and juridical structures persisted, including the Roman tenurial system as reworked by the Franks, which was based on allodial property.[16] Then, in the tenth century, the small allod came under attack, as a land market, stimulated by urban money and the growth of monastic estates, came into being. Beginning in the 980s, allod holders experienced downward social mobility which stimulated donations and sales to church establishments and the conversion of many small allods into precarious tenure as their peasant owners slipped towards dependence.[17] The feudal revolution of the early decades of the eleventh century forced the practical merger of the rural slaves and most of the free peasantry that theorists of feudalism had reckoned an earlier phenomenon.

The growth of the agrarian economy, by the same token, also preceded the inception of feudalism: the water mill, whose diffusion was presumed by many historians to have been stimulated by feudalism as greedy lords sought to extend this lucrative monopoly,

14 Guy Bois, *The Transformation of the Year One Thousand* (Manchester, Manchester University Press, 1992), p. 60.

15 *Ibid.*, Ch. 1, *passim.*

16 The fiscalists would agree that the fact that *domini* collected taxes from peasants did not alter the allodial nature of the latter's tenure.

17 Bois, *Transformation*, pp. 45–52.

was firmly in place in the ninth and tenth centuries, having been a cornerstone of the agricultural expansion which Bois identifies, along with three-course rotations and the heavy plough, as innovations introduced by the free peasantry.[18] Bois's view of the transition, then, is a synthesis of two processes: the disintegration of the State and ancient fiscality, which developed from the top down, and a parallel process of reintegration, from the bottom up – 'the first sign of it being the enlargement and consolidation of small peasant production on the technical and economic level'.[19] Feudalism, in turn, was also – somewhat paradoxically in view of stereotypes long accepted – 'the age par excellence of small-scale production', every one of whose structures was designed to 'buttress, protect and perpetuate' such production.[20] It was the economic and technological success of the small allod holder of the eighth to the tenth centuries, the peasants who brought about the agrarian revolution of the early Middle Ages, that made their domination so attractive to the incipient class of feudal lords. This explains, in part, why feudal society was – as Bois pictures it – economically aggressive and innovative, as lords were able to foment the creation of urban demand for rural agricultural surplus, to commercialise agricultural production, and to stimulate the secondary and tertiary functions of towns.[21]

I find Bois's innovative picture of an economically dynamic feudalism most compelling, especially for that area of medieval Europe with which I am most familiar, eastern Spain, where (in the case of Valencia), from the conquest of the mid-thirteenth century on, feudal lords coexisted with a strong urban commercial bourgeoisie linked together by a highly dynamic, highly commercialised agricultural system.[22] One prerequisite for such a system is a peasantry fully possessed of the technological wherewithal to power the agricultural sector. Thus I am convinced that Bois has got the

18 *Ibid.*, pp. 97, 113.

19 *Ibid.*, p. 117.

20 *Ibid.*, pp. 17–18. This view is adopted for thirteenth-century Valencia by Antoni Furió and Ferran Garcia, 'Dificultats agràries en la formació i consolidació del feudalisme al País Valencià', *Estudi General*, 5–6 (1985–86), 291–310, on p. 298. They speak of the hegemony of small peasant exploitation there, with very few seigniorial reserves, and the persistence of small allods. This view supports my notion of the importance of cross-cutting solidarities, particularly in the social organisation of irrigation in Valencia, where the same peasant might function as a villager, a member of an irrigation community, and a craft guild all at the same time.

21 Bois, *Transformation*, pp. 86–8.

22 Cf. Ferran Garcia Oliver, *Terra de feudals* (Valencia, Institució Valenciana d'Estudis i Investigació, 1991), esp. p. 112.

sequence of technological change correct, with respect to social movements: the agrarian revolution was closely associated with the economic vitality of small allodial peasants, empowering conjugal units, weakening extended families.[23] This is the basic process of 'reintegration' that accompanies, *pari passu*, the final demise of 'ancient structures', in particular rural slavery and the particular mind-set that its persistence – however numerically insignificant – generated among the 'masters',[24] and of the last remnants of ancient fiscality.

Bois's analysis provides a number of concrete criteria for the transition, particularly its final phases: how long did rural slavery last? When did the small allod holder become subject to a lord? When did a market in land begin? When did the city/country relationship show signs of change and urban markets emerge? When did the water mill cease being a motor of peasant expansion and become an instrument of seigniorial oppression?

Most of these questions had already been answered by Pierre Bonnassie with respect to Catalonia. For Bonnassie, in Catalonia of the ninth and tenth centuries, the 'ancient system' was still intact: there were public offices and no private fiefs, and, more importantly, it was still a 'slave society'. Not only had rural slaves survived, the political system appeared to require them: condemnation to slavery by public courts was still in evidence in the late tenth century and, as foreign supplies of slaves dried up, they were recruited ever closer to their place of servitude.[25] Slavery disappeared around the third decade of the eleventh century, but 'without any serfdom having emerged to take its place'. In Catalonia, at least, the new form of servitude did not emerge until around 1060.[26]

Evidence for the survival of rural slavery is also found in tenth-century Asturias, León, Galicia, and Castile, 'despite its reputation as a land of freedom'.[27] In the Asturo-Leonese kingdom there are very explicit ninth- and tenth-century documents demonstrating the persistence of rural slavery there: a bishop donating his *servos uel*

23 Bois, *Transformation*, p. 117.

24 *Ibid.*, p. 23: only an individual with slaves working his land was reckoned a 'person of distinction'.

25 Pierre Bonnassie, *From Slavery to Feudalism in South-Western Europe* (Cambridge, Cambridge University Press, 1991), pp. 36–7.

26 *Ibid.*, pp. 56–7. Bonnassie's scheme can be confirmed almost exactly in other areas, such as Auvergne: the last reference to public authority *c.* 970, the usurpation of the title of count by a viscount *c.* 985, multiplication of castles after 980, the last mention of *mancipia* around AD 1000, together with the emergence of local 'customs' (Lauranson-Rosaz, *Auvergne*, p. 405).

27 Bonnassie, *From Slavery to Feudalism*, p. 115.

ancillas to a monastery in 867, a family of *mancipes* donated to a monastery in 898; and then in the tenth century donations of *villae* with their men, both free (*ingenuis*) and unfree (*servis*).[28] In Galicia, documents of the same period reveal *mancipia* or (following the Visigothic formulary tradition) *pueri* and *puellae. Mancipia* were often confused with domestic *seruitales,* a term which, on occasion, might also refer to rural slaves. The same two terms could also refer to dependent peasants.[29] Sánchez-Albornoz recorded references to *casatos* in the county of Castile, persons who were unfree and whose work (*opus servile*) on seigniorial demesnes was more onerous than that of free peasants. Rural slavery, in fact, increased in Castile in the eleventh century.[30] In Galicia and elsewhere one finds *servi* paying both taxes and rent at the same time.

Bois's construction of rural slavery is highly congruent with (and somewhat derivative of) that of Bonnassie. Others are not so sure. Salrach, who formerly subscribed to Bonnassie's thesis, now demurs: the term *servus* may in fact have no longer meant 'rural slave'. Visigothic law which, as Bonnassie observed, is replete with mentions of rural *servi,* may have retained the juridical sense of the term, even as it had been transformed at the level of production.[31] Here the fiscalist paradox is excruciatingly clear: if terms like *servus* have only fiscal meaning, what can one say about peasant production?

Catalan allods for Bonnassie were under threat in the ninth and tenth centuries by absorption into ecclesiastical estates, even as they were in Bois's Lournand.[32] But in Catalonia, as in the other Christian kingdoms, the allodial base was always being renewed through assarting in frontier areas in the process called *adprisio* (*presura,* in Castilian). *Adprisio* was a kind of squatter's right, an inducement for peasants to improve their worth by settling on a frontier (or for lords to settle peasants there) on attractive terms. The southern expansion of the frontier set up a distinctive gradient

28 Claudio Sánchez-Albornoz, 'Los libertos en el reino astur-leonés', in his *Instituciones medievales españolas* (Mexico, UNAM, 1965), pp. 317–51 (examples cited on pp. 326, 332, 334, 335). He notes (p. 351) that slaves were most numerous in Portugal, fewer in León, and nonexistent in Castile (of course not, because his theory of free proprietors requires that there be none).

29 Isla Frez, *Sociedad gallega,* pp. 204–5, 207.

30 Claudio Sánchez-Albornoz, *Despoblación y repoblación del valle del Duero* (Buenos Aires, Instituto de Historia de España, 1966), p. 321 n. 100; and comment by José Angel García de Cortázar, *El dominio del monasterio de San Millán de la Cogolla (siglos X a XIII)* (Salamanca, Universidad, 1969), pp. 228–30.

31 Salrach, in *De la Antigüedad al Medioevo,* discussion on pp. 171–2.

32 Bonnassie, *From Slavery to Feudalism,* p. 116; Bois, *Transformation,* p. 52.

of freedom, once the feudal transition was achieved: as the frontier advanced, pressure on allods was increased in what had now become a hinterland.[33]

In Castilian history the status of the peasant has been a polemical issue. The doyen of medievalists, Claudio Sánchez-Albornoz, had insisted that the Castilian peasant of the high Middle Ages was a small, free proprietor. In the light of the analysis of allodial holdings by Bois and Bonnassie, however, the phenomenon is quite clear: of course the free proprietor was dominant in Castile in the eighth to the tenth centuries: he was everywhere! Thereafter one need only stipulate, in the case of Castile, that in specified frontier situations, the allod persisted and a more fuller feudalisation of such territories was delayed.

As for the nexus of market expansion, agricultural growth, and technological advance that Bois describes, the model fits Catalonia and no doubt the other kingdoms as well. Bonnassie is adamant on locating the most complex technological innovations precisely on the site of peasant allods, namely the *villae* of the high Middle Ages.[34] Small allod holders built the first grist mills, as well as the earliest irrigation systems, all antedating the completion of the transition – as we shall see later in this chapter.

The transformation of the *villa*

The decline of the Roman *villa* or, rather, its transformation into the *aldea* or *alquería* of high medieval Hispania, whether Muslim or Christian, is a conundrum. Magnou-Nortier notes that in the Narbonnaise, charters from the eighth to the tenth centuries always specify that the parcels described are in a *villa*. *Villae* were composed of some manses and a variety of agricultural parcels, but fields or vineyards are always described as pertaining to a *villa*, never a manse. The *villa* then has acquired 'juridical personality' (which Magnou-Nortier attributes to Roman tradition) and was also the centre of religious life, even before the parish structure was fully in place (the term *parrochia* does not appear in early medieval sources, but the villa assumed its function, if supplied with a church).[35]

33 Bonnassie, *From Slavery to Feudalism*, p. 116. Thomas F. Glick, *Cristianos y musulmanes en la España medieval* (Madrid, Alianza, 1991), p. 280.

34 Pierre Bonnassie, *La Catalogne du milieu du Xe à la fin du XIe siècle: Croissance et mutations d'une société*, 2 vols. (Toulouse, Université de Toulouse-Le Mirail, 1975), I, 215–22.

35 Elisabeth Magnou-Nortier, *La société laïque et l'église dans la province ecclésiastique de Narbonne de la fin du VIIIe a la fin du XIe siècle* (Toulouse, Université de Toulouse-

A *villa* for Magnou-Nortier, then, is an ensemble composed of the *villa* proper (designated by a place-name) and dependencies called *adiacentiae, apendiciae, pertinentiae,* etc., usually located in the same *territorium,* or administrative district. Medievalists well know that when *villae* were alienated, they were always defined by a stock-list of such *pertinentiae,* such as mills, vineyards, pastures, rights-of-way, and so forth. The presence of the same lists in donations of whole *and* partial *villae* alike suggests that what was being donated was the income from such *villae.* The King delegated their management to others, and then they slowly became the private property or eminent domain of some lord. The language of such charters constitutes evidence for Magnou-Nortier that the *villae* of the high Middle Ages could not have been great domains, where there must always be some distinction made between the reserve and tenancies.[36]

For Bonnassie too, the *villa* of the high Middle Ages could not possibly have been a great domain: had it been so, Catalonia would have been a country of *latifundia* which, clearly, it wasn't.[37] It was more likely a 'mosaic of peasant allods', under the fiscal direction of a master[38] (a local magnate who, in Bois's construction, was not yet a feudal lord but rather an aristocrat who had undertaken certain public functions, such as the collection of taxes and protection of peasants). *Villae,* then, were communities of free men which, in the course of the tenth century, were weakened as peasant allods fell in ever greater numbers into the hands of monasteries and nobles, only to see their 'franchise' destroyed during the 'feudal revolution' of 1040–60.[39] In the case of the Duero Valley (see below), García de Córtazar sees a double origin for the tenth-century *aldea*: some evolved out of the Roman *villa,* originally as settlements for servants (*casatos*) of estates and their descendants. Others, however, were of much more recent foundation, settlements by extended

---

Le Mirail, 1974), pp. 147–59. See below, p. 105, on the linkage between village and parish in the reorganisation of the countryside.

36 Elisabeth Magnou-Nortier, 'La gestion publique en Neustrie: Les moyens et les hommes (VIIe–IXe siècles)', in H. Atsma, ed., *La Neustrie: Les pays du nord de la Loire de 650 a 850,* 2 vols. (Sigmaringen, Jan Thorbecke Verlag, 1989), I, 271–318, on pp. 273–85: 'Villa, locus, praedium, possessio, res. Etude Lexicographique'. Although Magnou-Nortier's conclusions often seem exaggerated or reductionist, the kind of close analysis of terminology that she favours is absolutely necessary.

37 Cf. Miquel Barceló's observation that subsequent to the break-up of Roman estates in the third and fourth centuries, there were *no* settlements which weren't small (comment in Antonio Malpica Cuello, *La cerámica altomedieval en el sur de Al-Andalus* (Granada, Universidad, 1993), p. 142).

38 Bonnassie, *La Catalogne,* I, 215–22.

39 Bonnassie, *From Slavery to Feudalism,* pp. 250–1.

families.[40] According to Sánchez-Albornoz, *villa* place-names in northern Spain are proof of internal colonisation and of the persistence of a stable population in the northern mountains.[41] Nevertheless, it is difficult to interpret descriptions like that of the *villa de Massella – iacebat erema sine tectos et sine hominos* ('it lay barren without roofs and without men') – or another tenth-century *villa, cum omnia sua prestantia vel adjacentia* ('with all its appurtenances and dependencies').[42] Are these the ruins or successors of formerly Roman *villae*, or of further evolved settlements of Visigothic times more along the lines of medieval peasant villages?

Villae novae, of course, leave no doubt as to their origin, which Gautier-Dalché associated with the process of *presura* by extended family groups.[43] The extended family consisted of a kinship group headed by a man who represented all the *herederos*, that is, his siblings and children considered as a corporate descent group. Women also figured in lists of *herederos*; thus there was still cognate, as well as agnate, filiation. In some of these groups, kinship terminology appears to have the sense of 'associate' and not of blood relatives (something like what Crone refers to, in the case of similar groups in Islamic society, as representing the 'after-image' of tribal organisation). In any case, the *aldea* emerged just as the extended family was yielding in significance to the conjugal pair. Village families joined together in a *concilium*, a village assembly with executive power.[44] Wherever there was a significant amount of migration, as in the settlement of a frontier area, extended families were bound to pare down to conjugal units.

The definitive emergence of the *aldea* of the high Middle Ages out of the late Roman/Visigothic *villa* also seems to have been a phenomenon of the early eleventh century. Portela and Pallares call the Gallegan *villa* of the ninth and tenth centuries a '*pre-aldea*'. The social structure of their *pre-aldea* is what Bois predicts: within

40 Julio Angel García de Cortázar, 'Del Cantábrico al Duero', in García de Cortázar, ed., *Organización social del espacio en la España medieval* (Barcelona, Ariel, 1985), pp. 43–83, on pp. 69–71. See also Julio Valdeón Baruque, 'Señores y campesinos en la Castilla medieval', in *El pasado histórico de Castilla y León*. Vol. I. *Edad Media* (Burgos, Junta de Castilla y León, 1983), pp. 59–86, on p. 69. In the case of foundation by extended families, there is an analogy with the clan-based *alquerías* of Al-Andalus.

41 Sánchez-Albornoz, *Depoblación y repoblación de la Valle del Duero*, pp. 269–70.

42 *Ibid.*, pp. 275, 277.

43 Jean Gautier-Dalché, 'Reconquête et structures de l'habitat en Castille', *Castrum* 3, pp. 199–206, on p. 202.

44 Reyna Pastor, *Resistencias y luchas campesinas en la época del crecimiento y consolidación de la formación feudal. Castilla y León, siglos X–XIII* (Madrid, Siglo Veintiuno, 1980), pp. 20–8.

the *villa*, aristocrats dominated a large group of small and middling free proprietors plus a numerous group of rural slaves.[45]

Incastellamento

Incastellamento is the name given by Pierre Toubert to a process of the general reorganisation of the countryside of Lazio in the high Middle Ages around AD 1000 which resulted in the disappearance of dispersed habitats and a regrouping of settlements around castles. After the reorganisation was completed by the middle of the eleventh century, there was no place that did not depend on a castle.[46] Early castles were built on uninhabited, elevated sites, chosen for their defensive positions and typically called *rocca*, rock, in the documents. The purpose of these castles was not to bring security to the countryside, as historians had generally assumed, but to dominate it, to enforce the feudal 'ban' on the now servile populations it held, and to defend against neighbouring castles.[47] The charters show a number of processes taking place simultaneously around the end of the tenth century: the building of castles, the gathering together of settlements (*congregatio populi*), the consolidation of dispersed dwellings (*consolidatio fundi*), the end of rural slavery and concomitant loss of freedom by small allodial peasants. Because the Church was the driving force behind the *consolidatio fundorum*, in order to concentrate holdings built up by donation and purchase on an *ad hoc* basis, and because such concentration was made easier with a servile peasantry, Toubert reckons bishops and abbots as the oldest and possibly the principal promoters of *incastellamento* in its broadest sense.[48]

But local village churches were also prominent in the general reorganisation of the countryside. *Incastellamento* was inevitably accompanied by the consolidation of parishes around village churches, many of which arose in the tenth century from castral chapels.

45 Ermelindo Portela and María Carmen Pallares, 'De la villa altomedieval a la fortaleza del siglo XV. Fuentes escritas y arqueología en Galicia', in *Coloquio Hispano-Italiano de Arqueología Medieval* (Granada, Patronato de la Alhambra, 1992), pp. 215–21, on pp. 218–19.

46 Pierre Toubert, *Les structures du Latium médiéval* (Rome, Ecole Française de Rome, 1973); I have used the Spanish version which excerpts from that longer work the portions dealing with *incastellamento* and the feudal transition: *Castillos, señores y campesinos en la Italia medieval* (Barcelona, Crítica, 1990), pp. 208, 211.

47 On this point, see André Debord, 'The Castellan Revolution and the Peace of God in Aquitaine', in Richard Landes and Thomas Head, eds.,*The Peace of God: Social Violence and Religious Response around the Year 1000* (Ithaca, Cornell University Press, 1992), pp. 135–64, on p. 144.

48 *Ibid.*, pp. 206 n., 210, 266, 281–2; Bonnassie, *From Slavery to Feudalism*, p. 212.

Village cemeteries preceded the establishment of the church, and villages took shape in sites fixed in the first place by the location of cemeteries. The labours of the living, in Fossier's vivid image, assembled around their dead.[49] Fossier's construction of the dynamics of early village organisation is made even more compelling by the possibility of testing the cemetery location hypothesis archaeologically. At the moment each village was equipped with its own church, as a natural pole of allegiance, peasants then became identified with specific settlements and these, in the course of the tenth and eleventh centuries, evolved recognised boundaries, sometimes coextensive with recently-established castral districts.[50]

In Catalonia, the spatial reorganisation took place with shocking speed and violence, thanks to a civil war in the County of Barcelona provoked by Mir Giribert and his barons. The transition was completed in one generation: the first mentions of homage date to 1020; the process was completed by 1060. Rural slaves disappeared, the better off of the allodial peasants joined forces with the nobility; the remaining peasants lost their freedom.[51] As for the system of castles that emerged, 48 per cent of the 800 castles that were in existence in 1350 had been built before the end of the eleventh century.[52] In New Catalonia, *incastellamento* was associated with settlement of the plains, particularly in the tenth century, and was accompanied by a restructuring of settlement and economic activity in the new castral zones, which included settlements in the immediate vicinity of castles, plus dispersed, smaller settlements on the top of hills within the castral districts.[53]

In its full form, in the mid-fourteenth century the Catalan

49 Robert Fossier, *Enfance de l'Europe*, 2 vols. (Paris, Presses Universitaires de France, 1982), i, 192–4. In Catalonia, as early as 833, some *castra* far from nuclei of settlement had their own chapels; Manuel Riu Riu, 'L'Aportació de l'arqueologia a l'estudi de la formació i expansió del feudalisme català', *Estudi General*, 5–6 (1985–86), 27–45, on p. 44.

50 Chris Wickham, *The Mountains and the City: the Tuscan Apennines in the Early Middle Ages* (Oxford, Clarendon Press, 1988), pp. 176, 336.

51 Bonnassie, *From Slavery to Feudalism*, pp. 108, 156–7.

52 Philippe Araguas, 'Le réseau castral en Catalogne vers 1350', *Castrum* 3, pp. 113–22, on p. 120. Fifteen per cent of the total were built in the tenth century or before, including some in hilltop sites, reoccupying *oppida* in the ninth and tenth centuries, particularly in the Pyrenees or pre-Pyrenees (Savassona, Ullastret, Sant Martí d'Empúries, Olèrdola, etc.). See Xavier Barral i Altet, 'Quelques exemples d'habitat groupé en hauteur en Catalogne (Xe–XIe siècles)', *Castrum 2*, pp. 85–96.

53 Jordi Bolós, 'Fortificacions fronteres situades entre els rius Anoia i Gaiá. L'estructuració d'un territori el segle X', *II CAME*, II, 113–22; Francesc Fité, 'Arquitectura militar y repoblación en Catalunya (siglos VIII al XI)', *III CAME*, I, 193–235, especially p. 206.

castle network can be divided into four zones (from greatest to least density of castles): a central zone (Bagès/Anoia, Conca de Barbera), the north-east (Girona/Empordà), north centre ('Old Catalonia'), and south (see Map 7).[54] As is true of the Andalusi castle systems studied by Guichard, the Catalan castle system was not primarily defensive in nature: it was most dense in the middle of the country, the most protected part; the southern frontier, which in the eleventh and twelfth centuries certainly was exposed to Muslim incursions, was the least defended zone. Only a dense node of castles around Lleida could be reckoned as having a pre-eminent defensive role. Interestingly enough, fully 49 per cent of Catalan castles were controlled by church corporations (military orders and monasteries).

There were numerous *castra* or *castellos* in the Christian kingdoms of the tenth century which were wholly rural and which protected dependent *villae*.[55] Both Catalonia and Castile take their names from castles, or rather from their custodians, the *castlà* (> *catalá* = Catalan), or *castellano*. But in Castile, to the north of the Duero at least, castles were abandoned in the eleventh century just when, in Catalonia, *incastellamento* entered a phase of rapid growth.[56] The feudal reorganisation there took place around fortified towns, like Burgos.

All of the *incastellamento* theorists (Guichard, Bazzana, Cressier, Barceló, etc.) have developed a view of feudalism that is highly conditioned by their view of Al-Andalus as a non-feudal or tributary society. Scales, however, argues that a Visigothic-style feudal system prevailed in eighth-century Al-Andalus, rather than the eastern *iqta*,[57] thanks to a deal brokered by the Gothic Count Ardabast. Besides a much higher rent (two-thirds of the produce, instead of the tenth that is statutory in Islamic law), other Western features – such as the rendering of homage with a hand-clasp – intruded. This system had a number of interesting social consequences: it weakened tribalism among Arabs by providing an incentive for them to associate strongly with their estates. Furthermore, and as a consequence, this explains why the Marwanids (the extended family of the Umayyad

54 Araguas, 'Réseau castral', pp. 114–15.

55 Luis García de Valdeavellano, *Sobre los burgos y los burgueses de la España medieval* (Madrid, Real Academia de la Historia, 1960), p. 53.

56 Bonnassie, *From Slavery to Feudalism*, p. 212. Jean Gautier-Dalché, 'Châteaux et peuplements dans la Péninsule Ibérique (Xe–XIIIe siècle)', in *Châteaux et peuplements en Europe Occidentale du Xe au XVIIIe siècle* (Auch, Centre Cultural de l'Abbaye de Flaran, 1980), p. 101.

57 The *iqta* is typically portrayed as a purely fiscal fief which implied no loss of freedom by rentpayers.

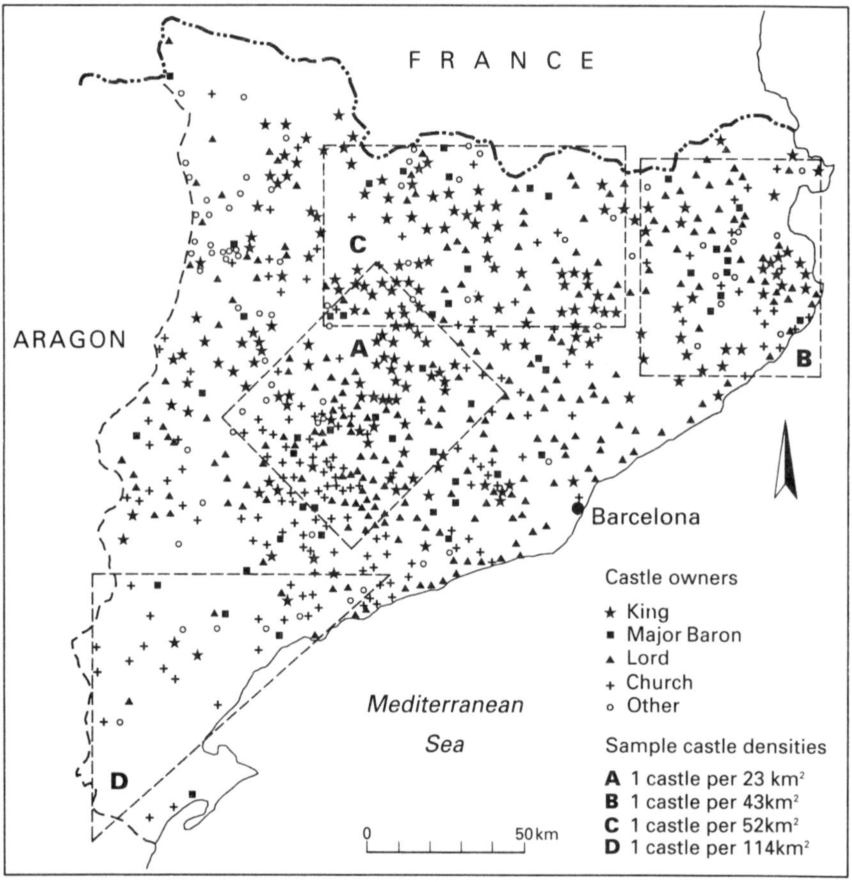

Map 7 *Incastellamento in Catalonia*

emirs and caliphs), as an interest group in Cordoban politics, were so numerically insignificant during the *fitna* of the eleventh century. In spite of the documented fecundity of the Umayyads, the number of identifiable family members kept diminishing. The reason was their attachment to their fiefs and, as a possible consequence, intermarriage with local Muwallad families.[58]

Nevertheless, fiefs alone do not make a feudal system of social organisation or of governance. Although in Al-Andalus, as in the Islamic east, military service was sometimes rewarded with fief-like grants, and although there were aristocrats who lived off their rents (as there are in practically all societies), there was no feudalism because the aristocratic class did not live off the rents of an ascriptively dependent peasantry, and the State was supported preponderantly on taxes. To invoke 'Wickham's Razor', the balance never shifted from taxes to rents there.

The discussion of *incastellamento* begun by Toubert's book and by a meeting on feudalism directed by Toubert and held in Rome in 1978[59] set off waves of historiographical re-evaluation wherever there were medieval castles. The oddity of the Spanish discussion is that the vast majority of *incastellamento* studies carried out in the 1980s had to do with Muslim, that is, non-feudal, castles, the results of which are discussed in Chapter 2. This state of affairs provoked an interesting comment by Toubert in 1988.[60] These fortified habitats played a role that exceeded a simple military one, extending to the organisation and structuring of rural life. He called for a more finely-tuned periodisation of the process, a general typology of castles and systems of castles, and an examination of the relationship between castles and their dependent villages. By then it had become abundantly clear that the same methodology could be applied both to feudal and non-feudal castle systems: both *castra* and *ḥuṣūn* had been called into being in the ninth or tenth centuries to assure 'public' control of a first phase of rural growth and settlement. Indeed, one of the chief characteristics of castles was

58 Peter C. Scales, *The Fall of the Caliphate of Cordoba: Berbers and Andalusis in Conflict* (Leiden, E. J. Brill, 1994), pp. 117–19, 122–30. This dynamic is similar to, but not as compelling, as that proposed by Acién Almansa (Ch. 3, above).

59 *Structures féodales et féodalisme dans l'Occident Méditerranéen* (Rome, Ecole Française, 1980); Spanish version, *Estructuras feudales* (n. 3, above). Since the 1978 meeting, a number of important symposia were held to present new research on *incastellamento*. See, in particular, the published acts: *Flaran 1: Châteaux et peuplements en Europe Occidentale du Xe au XVIIIe siècle* (Auch, Abbaye de Flaran, 1980); *Habitats fortifiés et organisation de l'espace en Méditerranée médiévale*, ed. A. Bazzana, P. Guichard, and J. M. Poisson (Lyon, Maison de l'Orient, 1983); and *Castrum 2*.

60 Pierre Toubert, 'Préface' to André Bazzana, Patrice Cressier and Pierre Guichard, *Les châteaux ruraux d'Al-Andalus* (Madrid, Casa de Velázquez, 1988), pp. 9–13.

their functional polyvalence: they had strategic, economic and administrative roles. There has been some discussion of castles as *central places,* that is, as sites regularly appearing in the countryside which serve a topographically defined zone in a variety of organisational and service capacities.[61] That means that

> At a certain level, (*incastellamento*) is a method of mental organisation: castles and/or central nuclei . . . can be seen as nodes for the interaction of a specific group of political and economic, as well as religious, processes: they are, that is, an organisational expedient which permits historians to conceptualise these processes in a dialectical relationship. Other concepts or institutions might in principle have the same scope, but few function with the same efficacy.[62]

The process has been little studied in the rest of Christian Spain. In Galicia, there was a reorganisation of settlement in the eleventh century, with *villae* characteristically coming into the orbit of monasteries. But, in spite of the multiplication of references to castles in the eleventh and early twelfth centuries, Frez does not see *incastellamento* in the sense intended by Toubert. Rather, there was a sociopolitical reorganisation, whereby old *villae* were converted into new jurisdictional units, called *territorios,* under the *mandamentum* of a lord.[63] García de Cortázar and Diez Herrera describe the progressive politicisation of space in the Cantabrian region of the eleventh century, characterised by the spontaneous implantation of a society grouped in small nuclei, for which a new noble class sets a code of rights and duties[64] – *congregatio populi* – but apparently without castles.

A similar reorganisation occurred in Castile and León from approximately the mid-twelfth century to 1230. There, castral units called *tierras* are documented in the late twelfth century.[65] In the eastern Duero Valley, Escalona describes a process of rural reorganisation accompanying the rise of feudalism in the eleventh century, whereby original dispersed settlements crystallised in a single

61 See comment by R. Comba and ensuing discussion in *Habitats fortifiés* (n. 59, above), p. 192.

62 Chris Wickham, 'L'incastellamento ed i suoi destini, undici anni dopo il *Latium* de P. Toubert', *Castrum 2,* pp. 411–20, on p. 414.

63 Isla Frez, *Sociedad gallega,* pp. 244–6.

64 José Angel García de Cortázar and Carmen Diez Herrera, *La formación de la sociedad hispano-cristiana del Cantábrico al Ebro en los siglos VIII a XI* (Santander, Estudio, 1982), pp. 53–4.

65 Cristina Jular Pérez-Alfaro, '*Alfoz* y *tierra* a través de documentación castellana y leonesa de 1157 a 1230. Contribución al estudio del *dominio señorial*', *Studia Histórica. Historia Medieval,* 9 (1991), 9–42, on pp. 33–4. Jular notes, however, that the *tierra* or *tenencia* was not always coextensive with the jurisdiction of a castle.

village (*aldea*), served by a single church.[66] Yet, if there was no systematic *incastellamento,* we must still know how force was deployed spatially to effect the reorganisation of the countryside that feudalism implies.

Both Christian and Muslim castles appear to have their remote origins in the process of abandonment of lowlands and reoccupation of higher emplacements that had been abandoned in imperial times.[67] Of course this earlier phase, when elevated village sites and *oppida* were resettled and fortified – a process described in Chapters 1 and 2 – was common throughout the Mediterranean world in late Roman times and represents a process of *incastellamento*: habitats, as Wickham notes, 'are always crystallizing and decrystallizing in *all* centuries, especially in the Mediterranean and . . . differently in each microregion'.[68] Thus it is not surprising that similar processes should have been set in motion, for similar reasons, in Christian and Muslim societies both in the *incastellamento* of our first transition and in that of the eleventh (really, ninth to the eleventh) centuries. In part, the parallelism in this latter phase was a reflection of the process of growth in the agrarian sector in the entire Mediterranean world, stimulated by a technological revolution.

Fossier provides both a chronology and a geography of *incastellamento.* It begins in the tenth century in northern Italy, spreads to southern Italy at the beginning of the eleventh; in Provence it begins before 950, culminating around 1030; in Catalonia, it begins with the descent of warriors to the plains, 960–1020; in Auvergne, around the end of the tenth, and the beginning of the eleventh; in the Toulousain, from 1030 to 1050. Farther north, in the region between the Loire and the Rhine, the entire process unwinds around 100 years later.[69] This sequence raises the issue of the diffusion of this particular style of economic, social, and political reorganisation. It is unlikely that it arose spontaneously, autochtonously. But the means and agencies of diffusion have yet to be studied, or even identified. Thus García de Cortázar and Diez Herrera ask whether forms of social pressure that appear in tenth-century Castile might not have been imported from Navarre.[70]

As for the geographical distribution of distinct styles of *incastellamento,* Fossier distinguishes between two southern European

66 Julio Escalona Monge, 'Población y organización territorial en el sector oriental de la cuenca del Duero en al alta Edad Media', *III CAME,* II, 448–55, on p. 452.
67 Toubert, *Castillos, señores y campesinos,* p. 184.
68 Wickham, 'Incastellamento ed i suoi destini', p. 417.
69 Fossier, *Enfance de l'Europe,* I, 107–211.
70 García de Cortázar and Diez Herrera, *Sociedad hispano-cristiana,* p. 65.

zones, the first embracing Italy, Provence, and southern Aquitaine, characterised by the extension of *consuetudo castri* to regrouped lands, with a walled castle, the houses in a ring around it. This type has a Provençal sub-variety, characterised by villages with a castle at one end with houses grouped at its foot, not in rings, but in parallel streets ending in a gate. The second variety includes Catalonia, the Languedoc plain, and the Auvergne, from the Rhône to the Ebro. Here the settlements are not as concentrated: a *castrum* controls various *villae*.[71]

There is a nice chronological fit between *incastellamento*, on the one hand, and the 'ancient systems' arguments leading up to it, on the other. But I have been careful to describe the Fiscalist model as well as the Bois/Bonnassie alternative as heuristic ones which are useful in sorting out and tracking important variables. But they are not explanatory models, because I don't for an instant believe that the society so characterised (from the Visigoths through the Spanish kingdoms of the tenth century) can be described as 'ancient', nor can the disarticulated, residual features characterised by historians as ancient be considered a 'system' or 'structure'. A quite different reality prevailed beneath the 'spray of old words', to use a trenchant expression of Toubert's.[72]

Thus Isla Frez, following Bloch in preference to Bois, is reluctant to view the high medieval *servus* as a classical slave. That ancient slavery should have persisted in the Duero Valley or even Galicia is an exaggerated view, particularly inasmuch as many of the slaves documented there in the tenth and eleventh centuries were Muslim captives whose economic role was, in any case, exiguous. Similarly, Minguez doubts that such slavery as survived can legitimately be styled 'ancient' when every other element of the ancient system of production, such as the latifundio and an agricultural economy in which the slave was the main producer, had disappeared. The existence of rural slaves who have lost their productive function does not justify defining an entire society as ancient.[73]

71 Fossier, *Enfance de l'Europe*, I, 214–18. Fossier's scheme is important because regional analysis in medieval Europe is undertheorised. Italy and Provence fall within two distinct zones of agricultural technology, established by the terminology of harnessing: Richard Bulliet, *The Camel and the Wheel* (Cambridge, MA, Harvard University Press, 1975), pp. 204–5. Bois is correct in his judgement that the reorganisation of rural economies was pushed by technology. Therefore the relationship between different styles of technology, particularly those related to horse-power, and the process of reorganisation that we call *incastellamento* requires explanation. Castile and Portugal fall within the Italian group, which may explain their difference from Provence and Catalonia.

72 Toubert, *Castillos, señores y campesinos*, p. 270.

73 Isla Frez, *Sociedad gallega*, pp. 203, 205, 207; José María Minguez, 'Ruptura social e implantación del feudalismo en el noroeste peninsular (siglos VIII–X)', *Studia*

Societies and their social institutions are constantly recontextualised, and change recursively through the very act of repetition of familiar events.[74] Of the immediate pre-feudal societies of western Europe, it is sufficient to say that its institutions were no longer appropriate to, or were out of phase with the society. When that occurred, a major restructuring was called for. Whatever value-judgement one might want to place on the feudal transition, such major shufflings of the deck are usually functional. In this sense, the feudal reorganisation was functional in a number of ways. Given the failure of large-scale institutions, handing over small units of government to private persons operating in very localised areas was a good stratagem for certain objectives, such as keeping the economy going or, if Bois is right, restarting it. Peasants lost their freedom, but gained secure tenure, a trade-off that could be valued positively, at least in retrospect.

The transformation of the frontier

A fascinating aspect of Toubert's construction of settlement in medieval Lazio is that the processes described resemble those described for Spain, some of which have been misunderstood or misconstrued. In the famous debate over whether, and to what extent, the Duero Valley was 'depopulated' or devoid of settlement in the century or so after the Muslim conquest, Sánchez-Albornoz argued that Alfonso II had in fact taken the remaining population northwards and made the entire valley an empty buffer zone between the opposing blocs. Now however, both the study of place-names and some complementary archaeological evidence have demonstrated that this zone, which ran alongside the Andalusi middle march (*Thagr al-awsaṭ*), was in fact occupied by partially Arabised Christian settlements. First, there is no other way to explain pockets of Christians with Arab names in places, like Galicia, not noted as targets of settlement by Mozarabs fleeing Al-Andalus.[75] Place-name evidence yields similar conclusions. Angel Barrios and Luis Miguel Villar both argue that so many Roman and pre-Roman place-names could not have survived across a total depopulation

---

Histórica. Historia Medieval, 3:2 (1985), 7–32, on pp. 9–12. Minguez also points out that after 711, there was no semblance of public administration either in Cantabria or anywhere in the Central System (*ibid.*, p. 11).

74 Glick, *Cristianos y musulmanes*, p. 21; 'History and Philosophy of Geography', *Progress in Human Geography*, 11 (1987), 405–16, on p. 407.

75 See, e.g., Richard Hitchcock, 'Arabic Proper Names in the Becerro de Celanova', in *Cultures in Contact in Medieval Spain: Historical and Literary Essays Presented to L. P. Harvey* (London, King's College, 1990), pp. 111–26.

and that the 'Mozarab' names that turn up in tenth-century documentation were not of immigrants from Al-Andalus nor of Mozarabs moving southwards from settlements originally located to the north, but rather Christians who had never abandoned their settlements. Thus they see a diffuse Muslim/Christian frontier in the mid-tenth century, with some colonisation from both North and South, persistence of a rural population in the high villages of the central sierras and pockets of Muslim settlement interdigitated with Christian settlements.[76] Ramón Menéndez Pidal, Orlando Ribeiro and others argued that these zones were not depopulated and that terms like *desertus et incultus locus* referred to places lacking in administrative organisation. Therefore the sense of '*poblar*' is not 'to people' but to organise an area administratively.[77] Toubert shows that the same processes, described by the same terminology, occurred in Lazio, where the *populator* was the lord who directed the *congregatio populi* or *casamento,* which involved not only gathering people together (*far gente*), but also organising the cultivation of previously uncultivated territory.[78] Both sides of the Duero were fortified, and the relationship of these castles to settlement remains to be determined. Interpretation of these systems of castles (the Christians' an offensive system, the Muslims' defensive) is no doubt too narrowly conceived in military terms.[79]

When the southern bank of the Duero – the primitive Extremadura – was settled in the last quarter of the eleventh century, there took place a different kind of *incastellamento* linked to the new

76 Angel Barrios García, 'Toponomástica e historia: Notas sobre la despoblación en la zona meridional del Duero', in *Estudios en memoria del Profesor D. Salvador de Moxó* (Madrid, Universidad, 1982), I, 115–34; *Estructuras agrarias y de poder en Castilla. El ejemplo de Avila (1085–1320),* 2 vols. (Salamanca, Universidad, 1983), pp. 107–22; and 'Repoblación de la zona meridional del Duero. Fases de ocupación, procedencias y distribución espacial de los grupos repobladores', *Studia Historica (Historia Medieval),* 3:2 (1985), 33–82, on pp. 59–60. Luis Miguel Villar García, *La Extremadura castellano-leonesa: Guerreros, clérigos y campesinos (711–1252)* (Valladolid, Junta de Castilla y León, 1986), pp. 49–60. For discussions by Arabists, see Scales, *Fall of the Caliphate,* pp. 182–200, and Eduardo Manzano, *La Frontera de al-Andalus en época de los omeyas* (Madrid, CSIC, 1991), pp. 171–5.

77 See discussion by Abilio Barbero and Marcelo Vigil, *La formación del feudalismo en la Península Ibérica,* 4th ed. (Barcelona, Crítica, 1986), pp. 226–7. There is also some archaeological evidence pointing towards continuity of settlement in the Duero Valley: see, for south-eastern Galicia, Jorge López Quiroga and Mónica Rodríguez Lovelle, 'Una aproximación arqueológica al problema historiográfico de la "despoblación y repoblación en el valle del Duero", s. VIII–XI', *Anuario de Estudios Medievales,* 21 (1991), 3–9.

78 Toubert, *Castillos, señores y campesinos,* pp. 196–8.

79 Scales, *Fall of the Caliphate,* pp. 184–200. But see Eduardo Manzano, *Frontera,* p. 59, where he argues that the Thagr was not a line of defence, properly speaking, because the fortifications were not necessarily in contact with the enemy.

settlement and to the frontier: here there was a special kind of *consolidatio fundi*, whereby *aldeas* in wide areas (*alfoces*) were grouped around fortified cities (Sepúlveda, Avila, Segovia, and so forth).[80] Then, the conquest of Al-Andalus stimulated localised waves of *incastellamento* in areas of sparse settlement. There, either the preexisting networks of *ḥuṣūn* were appropriated and population regrouped around them, or else new castles were built in accord with official policy. Therefore, in 1293, Sancho IV ordered the town council of Seville to build a series of castles 'around which to concentrate the scant, dispersed population of the region. This interesting process of *incastellamento* permitted the organisation of a territory which had been marginal up to that time and to establish positions facing the new frontier with Portugal.'[81] This kind of royally directed *incastellamento* we might consider a secondary phase of the process, whose political origins were quite different from that of the eleventh century, even if the result was the same.

Hydraulic development in the feudal transition

First, let us consider the rôle of hydraulic mills across the transition. Mills for grinding grain were one of the key elements in the agrarian revolution of the high Middle Ages, demand for which was stimulated mainly by the lords' practice of demanding feudal rents in grain and, selectively, by the introduction of the horse-drawn, wheeled plough and the concomitant changeover from the two- to three-course rotation, which produced the oats to feed the horses.[82] This revolution, as we have seen, was driven by small farmers, the holders of allods, free peasants. In Italy, as in the rest of western Europe, mills – some of them owned by allodial peasants – were rarely mentioned till the second half of the ninth century.[83] Many tenth-century documents from the Duero valley record donations or sales of collectively owned mills: Hazam sold to Sahagún his right to exploit two mills on the Cea river, 'in septem dies singulas oras' (937); in the same year, Hazreb and his wife gave to the same monastery their shares in two mills (one-fifth and one-sixth, respectively); in 943 Ambroni sold to Placino half of one mill and

80 Jean Gautier-Dalché, 'Reconquête et structures de l'habitat en Castille', p. 204.
81 *Ibid.*, p. 206; André Bazzana, 'Les structures: fortification et habitat', in his *Habitats fortifiés et organisation de l'espace en Méditerranée médiévale* (Lyon, Maison de l'Orient, 1983), p. 170; and Manuel González Jiménez, *En torno a los orígenes de Andalucía*, 2nd ed. (Seville, Universidad, 1988), p. 41.
82 According to Bois (*Transformation*, p. 115) both the three-course rotation and the water mill dated from Frankish times.
83 Toubert, *Castillos, señores y campesinos*, p. 252 n. 38.

one-eighth of another, and so forth. Collectively owned mills survived well into the next century: the monastery of Cardeña acquired one with no less than twenty-one proprietors in 1012, while throughout the eleventh century donations to that of San Millán, including one on the Arlanzón river from thirty owners, were of shares of mills, rather than entire ones.[84] In Catalonia, many mills were still owned by groups of peasant allod holders in the eleventh century: thus in 1064 Queno, his son and daughter, sold to another couple, Gilabert and Tota, four days and four nights in the third week of each month in the mill of Ametlla which they say had been theirs '*per aprisió*'.[85] In Castile–León as well as in Catalonia, these mills were horizontal, easy to build (as in the case of the Abbot Paul, who built one with his own hands in the ninth century), and typically owned by villagers who had shares (*vices*) in the mills and rights to its use for so many hours per week. Ownership in the mill and *vices* for use were distinct and the latter could be leased or sold without the owner forfeiting his ownership of the mill itself. In the course of the ninth and tenth centuries, monasteries bought up shares in these peasant-owned mills but generally without acquiring monopolies, because of the very abundance of mills.[86]

84 Sánchez-Albornoz, *Despoblación y repoblación del Valle del Duero*, pp. 285–6; Claudio Sánchez-Albornoz, 'Pequeños propietarios libres en el reino asturleonés. Su realidad histórica', in *Investigaciones y documentos sobre las instituciones hispanas* (Santiago, Editorial Jurídica de Chile, 1970), pp. 178–210, on p. 194; Salustiano Moreta Velayos, *El monasterio de San Pedro de Cardeña: Historia de un dominio monástico castellano (902–1338)* (Salamanca, Universidad, 1971), p. 167; García de Cortázar, *San Millán de la Cogolla*, pp. 87–8.

85 Francesc Fité i Llevot, 'Un apropament a l'estudi dels molins del Montsec i la Vall d'Ager', *Acta Historica et Archaeologica Mediaevalia*, 4 (1983), 207–38, on p. 224. Among recent historical and archaeological studies of Catalan horizontal mills, see, in addition, Jordi Bolós i Mascalans and Josep Nuet i Badia, *Els molins fariners* (Barcelona, Ketres, 1983); Francesc Español Bertran, 'Els casals de molins medievals a les comarques tarragonines. Contribució a l'estudi de la seva tipologia arquitectònica', *Acta Historica et Archaeological Mediaevalia*, 1 (1980), 231–54; Jordi Bolós i Mascalans and Angel Martínez i Huelde, 'El molí de la Torre Baldovina de Santa Coloma de Gramenet (Barcelonès)', *Acta Historica et Archaeologica Medievalia*, 7-8 (1986–87), 421–35; Jordi Bolós i Mascalans and Iñaki Padilla Lapuente, 'Un molí d'origen medieval: El Molinet de Naval', *Quaderns d'Estudis Medievals*, 1 (1980), 49–55; Jordi Bolós i Mascalans and Miquel Fàbregas i Sabater, 'Els molins de la conca mitjana del Llobregat durant l'alta edat mitjana. I. Introducció', *Quaderns d'Estudis Medievals*, 3 (1982), 556–68; Albert Virella i Bloda, 'Els molins d'aigua en l'alta medievalitat a ponent del Llobregat', *Miscel·lània Penedesenca*, 6 (1983), 249–71. These are mainly archaeological reports, with some documentation, of horizontal mills, most all of them variants of the *molí de cup* or *molí de rampa*. No Catalan vertical mill, to my knowledge, has been studied archaeologically.

86 Jean Gautier-Dalché, 'Moulin à eau, seigneurie, communauté rurale dans le nord de l'Espagne (IXe–XIIe siècles)', in *Etudes de Civilisation Médiévale. Mélanges offerts*

Gautier-Dalché, although his work on mills was influential, committed a major error by assuming that *aceña* was synonymous with *molino*. However, we now know that *aceña* referred to vertical mills, which, in Castile, were found on larger rivers, while the simpler horizontal mills (*molinos de rodezno*) were found on smaller streams.[87] Typically, horizontal mills were massed, giving rise to place-names like Río de los Molinos (*rivus molinarum*).[88]

Vertical mills, which required gearing and were expensive to build, were almost always owned by powerful people and linked, therefore, to the development of feudalism. They appear in documents as belonging to lords, particularly after AD 1000. In Catalonia, by the end of the tenth century most mill owners were lay aristocrats, references to fractional ownership in mills disappeared in the course of the eleventh century (indicating that peasants have lost control of milling), and subsequently church corporations replaced lay lords as the major owners of mills.[89]

The feudal mill, in contradistinction to the Andalusi mill, tended to be sited at the head of channels, with irrigation an afterthought (in Barceló's characterisation), with fields – *hortos subreganos* – (generally those of the miller only) irrigated from the *subtus rego*, the return ditch of the mill. The socially-based hierarchy of use implied was also reflected in the design of horizontal mills: the millrace (*balsa*) was continuous with the tank (*cubo*) that delivered the water to the wheel under pressure. In Andalusi mills, the two were separate and water could be diverted for irrigating terraces *before* it entered the mill.[90] I might add, that although Barceló's

---

à Edmond-René Labande (Poitiers, CESCM, 1974), pp. 337–49; Bonnassie, *From Slavery to Feudalism*, p. 249. See also the list of tenth-century collectively-owned mills in María José Carbajo Serrano, *El monasterio de los santos Cosme e Damian de Abellar* (León, Centro de Estudios e Investigación San Isidro, 1988), pp. 135–6 n. 60. On *vices* and their alienability, see discussion by Antonio Sáenz de Santa María, *Molinos hidráulicos en el Valle Alto del Ebro (s. IX–XV)* (Vitoria, Diputación Foral de Alava, 1985), pp. 210–13 and illustrative documents; and Gautier-Dalché, 'Moulin à eau', pp. 344–6.

87 Villar García, *Extremadura*, p. 335 n. 123: 'It is curious to observe on the map showing the distribution of documented *molinos* and *aceñas*, how the former are distributed on rivers corresponding to the secondary and even more so the tertiary hydrographic networks, while on the other hand aceñas are found on big rivers: the Duero and its principal affluents, corresponding – to judge by the material remains we can still see – to authentic milling complexes.'

88 José Angel García de Cortázar, 'El equipamiento molinar en la Rioja Alta en los siglos X al XIII', *Studia Silensia*, 3 (1976), 387–405, on pp. 397, 401.

89 Ramon Martí, 'Hacia una arqueología hidráulica. La génesis del molino feudal en Cataluña', in Miquel Barceló, ed., *Arqueología medieval: En las afueras del 'medievalismo'* (Barcelona, Crítica, 1988), pp. 165–94, on pp. 178–9.

90 *Ibid.*, pp. 169–70.

hypothesis of mill siting is controversial and must accommodate various anomalies, the basic point – that different mill types represent different ways of exploiting estates – has long been established for other countries, England, for example. There, horizontal mills were also built by groups of peasants and fell outside seigniorial bans, while the vertical mill, both more costly and more efficient, was preferred by feudal lords.[91] The intensification of seignorialism brought about by the Norman conquest of 1066 initiated a general technological reorganisation, whereby the least profitable horizontal mills disappeared and were replaced by vertical mills where there was sufficient water power, by windmills where there wasn't.[92] It is germane, therefore, to inquire whether, and under what conditions, feudalism may have set off the same cycle of events in Spain. Two varieties of horizontal mills, the *molí de cup/molino de cubo* [tank] and *molí/molino de rampa* [ramp], which both delivered water to the *rodezno* under pressure, proved to be highly functional wherever they were used (see Figures 1 and 2). The added pressure that it supplied made it possible to run horizontal mills with multiple millstones (as many as five), particularly when the mills were sited on irrigation canals and the head could be carefully controlled.[93] In England, by comparison, there were few or no mills with multiple millstones.[94] Therefore, in Valencia and other areas where horizontal mills were run on irrigation canals, there was no need for the transition and it did not occur.

91 Richard Holt, *The Mills of Medieval England* (Oxford, Basil Blackwell, 1988), pp. 120–2. Miquel Barceló argues that the advantage to lords of vertical mills was not necessarily their greater efficiency, but that they were an effective way to control the volume of peasant production.

92 Holt, *Mills*, pp. 112–13.

93 I first raised the issue of the horizontal to vertical transition in my article, 'Molins d'aigua a l'Horta de València medieval', *Afers* (Valencia), 9 (1990), 9–22. This article stimulated two responses, both of which made clear the more than adequate power and efficiency of the *molí de cup/ molino de cubo*: Sergi Selma Castell, 'Molins i rodes: entorn d'una discusió desafortunada', *Afers*, 15 (1993), 11–26, and Luis Pablo Martínez Sanmartín, 'La lluita per l'aigua com a factor de producció. Cap a un model conflictivista d'anàlisi dels sistemes hidràulics valencians', *Afers*, 15, pp. 27–44. See my response, 'Sobre la tipologia convencional dels molins hidràulics', *Afers*, 15, pp. 53–6. Luis Pablo Martínez Sanmartín also wrote a long excursus on horizontal versus vertical mills, suggested by the same considerations: 'Estructura social y cambio tecnológico. Una crítica a los determinismos tecnológico y economicista en la historia de la técnica', *Arbor*, 143 (1992), 103–31. On the *molino de cubo*, see also Nicolás García Tapia and Carlos Carricajo Carbajo, *Molinos de la provincia de Valladolid* (Valladolid, Cámara Oficial de Comercio, 1990), pp. 205–11. The *cubo* is the same as the Arabic *arubah*. Selma Castell, 'Molins i rodes', has good diagrams of the *molí de cup* (p. 22) and the *molí de rampa* (p. 24).

94 None at all, according to Holt, *Mills*, p. 131.

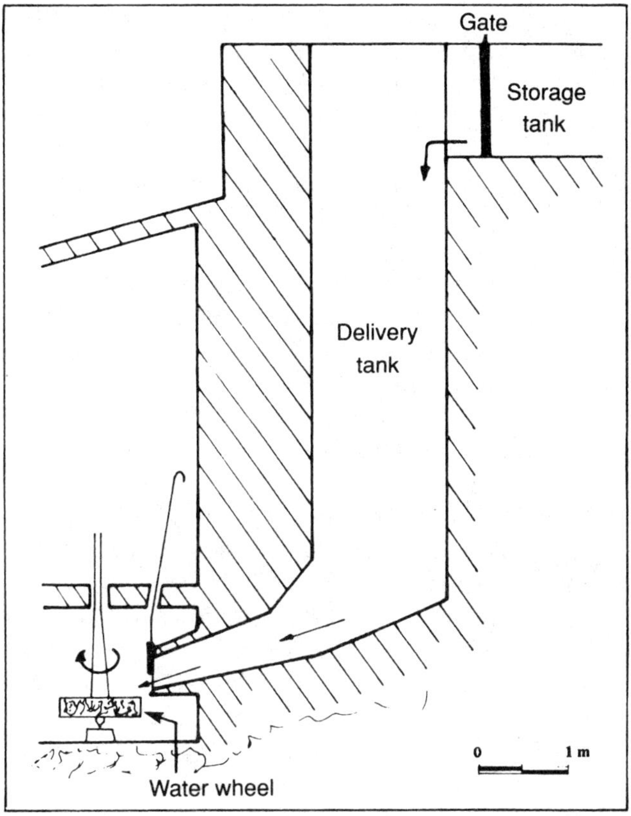

Figure 1 *Tank mill (molí de cup, molino de cubo)*

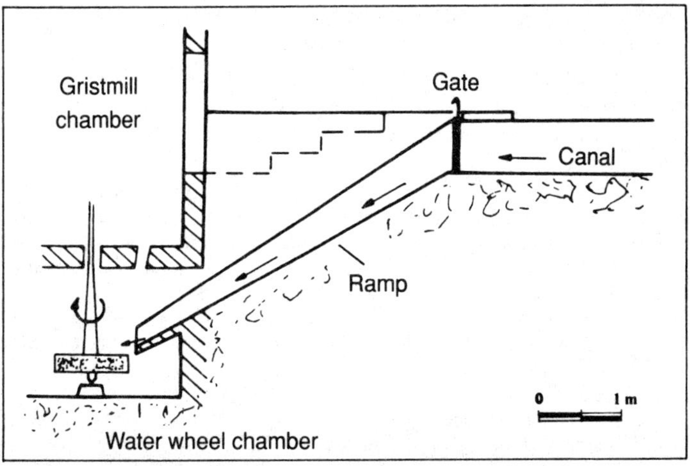

Figure 2 *Ramp mill (molí de rampa)*

The transformation is amply documented for Castile, however. In Burgos, horizontal mills were replaced in the course of the twelfth century by vertical ones, 'more permanent and efficient'.[95] The same was true of the area of northern Castile including Tordesilla, Valladolid, and Palencia, on the Duero river, where in the mid-twelfth century *aceñas* appear suddenly in the documentation where only *molinos* had been recorded before. Whereas in 1110, the abbot of Valladolid had seven mills on the Duero near Villavaquerín, at the beginning of the thirteenth century the thirteen mill sites on the same river between Villabañez and La Guarda had only vertical mills. The same was true of fourteen installations on the Pisuerga, from Reinoso up to the Duero confluence. *Molinos* continued to be recorded, but generally on small watercourses.[96] In Astorga too, *aceñas* on the Esla river in the early thirteenth century were designated 'new' while a horizontal mill on the Orbigo in the same period, owned by the same monastery, was 'old', leading Antón to suspect a similar transition.[97] There is even an instance recorded in Cuenca in 1185 in which the Bishop handed over some mills he owned on the Júcar river to be transformed into *aceñas*![98]

Mills, of course, constituted one of the classical seigniorial monopolies of feudal Europe. In the case of a 'banal' mill, peasants were obliged to mill at the lord's mill and in no other. Although banal monopolies on milling came to be the rule in Catalonia and, later in Valencia, in the crown of Castile the panorama was much more complex:

> With the exception of Aragón and Catalonia, in the crown of Castille and in Portugal, one finds a pattern in which peasants, bourgeois groups, and artisans – along with noble and ecclesiastical groups – own hydraulic mills. . . . The seignorialisation that Castille underwent in the middle ages . . . did not modify this model. The system of portions (*vezes* = *vices*) permitted monasteries, nobles and urban oligarchs to buy milling rights progressively without having to resort to the heavy expenses that the construction of a mill supposed.

95 Teófilo Ruiz, *Sociedad y poder real en Castilla* (Barcelona, Ariel, 1981), pp. 79–81.

96 Adeline Rucquoi, 'Molinos et *aceñas* au coeur de la Castille septentrionale (XIe–XVe siècles)', in *Les Espagnes médiévales. Aspects économiques et sociaux* (Nice, Faculté des Lettres, 1983), pp. 107–22, on pp. 108–11.

97 Isabel Alfonso Antón, *La colonización cisterciense en la meseta del Duero. El dominio de Moreruela (siglos XII–XIV)* (Zamora, Diputación Provincial, 1986), p. 168. Cf., for the transition in Provence, Henri Amouric, 'De la roue horizontale a la roue verticale dans les moulins à eau. Une révolution technologique en Provence?', *Provence Historique*, 33 (1983), 157–69. N.b. the siting of vertical mills on the 'canal de Sénas' (Marseillaise), p. 165.

98 Santiago Aguadé Nieto, 'Molino hidráulico y sociedad en Cuenca durante la edad media (1177–1300)', *Anuario de Estudios Medievales*, 12 (1982), 241–77, on p. 259.

Therefore it is not possible to 'establish a single system of ownership'.[99] The same lack of any seigniorial monopoly over mills is also reported for the Tierra de Campos.[100]

To sum up the role of mills in the transition to feudalism: there is good evidence that in Castile, the banal mill was the exception, not the rule. Because of the piecemeal way in which monasteries acquired mills, typically acquiring shares by donation (or by sale, disguised as donation), they rarely exercised the full seigniorial ban. One must therefore be careful when labelling mills as 'feudal', as many Spanish historians do today reflexively, without realising that there was more than one model as to what such a mill was. Rather than become preoccupied with legalisms of feudal jurisprudence, we should, as Miquel Barceló has urged, fix our sights on the nature of production. Here, it is clear that feudalism's emphasis on wheat as the prime medium of payment of rents was a prime driver of the system which led to the intensification and expansion of grain mills. Cortázar, too, is right, when he asserts that the grain mill, rather than the instrument of seignorialisation, was more the result of it.[101]

It is obvious that the Church was a major player in the expansion of milling, albeit a passive one – passive in that most of the mills that monasteries acquired were not built by them, but rather by lay lords or else by groups of allodial peasants. The eleventh century, the century in which the 'feudal system' was established, was also the great century of the proliferation of hydraulic projects.[102] We have noted that, in Castile and León, monastic establishments typically acquired mills, or, more typically, shares of mills during this century – mills which had been established by groups of allodial peasants. The result was that monasteries played a major role in the expansion of hydraulic energy, but a passive one: because of the way it obtained hydraulic power – in shares of mills – monasteries *treasurised* energy, holding it as an element of capital accumulation.[103] The monks were not necessarily interested in exercising

99 Hilario Casado Alonso, *Señores, mercaderes y campesinos. La comarca de Burgos a fines de la edad media* (n.p., Junta de Castilla y León, 1987), p. 199.

100 Pascual Martínez Sopena, *La tierra de campos occidental. Poblamiento, poder y comunidad del siglo X al XIII* (Valladolid, Institución Cultural Simancas, 1985), pp. 313–19.

101 García de Cortázar, 'Equipamiento', p. 405.

102 Robert Philippe, 'L'église et l'énergie pendant le XIe siècle dans les pays d'entre Seine et Loire', *Cahiers de Civilisation Médiévale*, 29 (1984), 107–17, on p. 107. See also David F. Noble, *A World without Women* (New York, Oxford University Press, 1993), p. 283, who suggests that priests, when forced to give up marriage, replaced 'woman power with water power'.

103 The notion of treasurisation of energy is that of Robert Philippe, *ibid.*, p. 112. In later centuries, cathedrals replaced monasteries as the main ecclesiastical mill owners: Villar García, *Extremadura*, p. 340.

banal monopolies: on the contrary, their objectives were more simple and practical: to be privileged clients of the mills so acquired was enough.[104]

It is clear that monastic management of mills increased significantly in the eleventh century, and the reason was the vast expansion of cereal culture.[105] In the twelfth century, the introduction of Cistercian monastic houses exaggerated the phenomenon. Cistercians typically acquired mills rights or rights to rehabilitate mills in disrepair, which they used for the advancement of their own estates. Because of their work ethic, the Cistercians used less servile labour than did other monasteries, increasing their dependence on mills, even though Cistercians tended to own mills to grind their own flour rather than acquire income from mills under their jurisdiction.[106]

Irrigation and feudalism

Early irrigation in Castile and León resembled milling, in that typically rights were expressed in shares held by allodial peasants. (The equivalence in some places of one *vicem* to one hour of milling was no doubt suggested by the universal tendency of irrigators to express water shares in hours, or fractions thereof.) There are many records of diversion of river water for irrigation and milling alike from the ninth and tenth centuries, when water could be taken by *presura,* contrary to the norm of Roman law, which regarded such water as public.[107] It was in order to defeat such appropriation that royally promulgated, Romanising law codes of the thirteenth century, like the *Siete Partidas* or the *Furs* of Valencia reiterated the public nature of flowing water.[108] Over the long run, however, it was difficult to seigniorialise irrigation systems because of the local control requirements that were discussed in Chapter 4. It is contrary to the

104 *Ibid.*, p. 113; Javier Pérez-Embid Wamba, *El Cister en León y Castilla* (Junta de Castilla y León, n.p., 1986), p. 109.

105 García de Cortázar, 'Equipamiento', p. 397.

106 Pérez-Embid Wamba, *El Cister en Castilla,* p. 108. See also Constance Brittain Bouchard, *Holy Entrepreneurs: Cistercians, Knights and Economic Exchange in Twelfth-Century Burgundy* (Ithaca, Cornell University Press, 1991), pp. 115–17; Constance Hoffman Berman, *Medieval Agriculture, the Southern French Countryside, and the Early Cistercians* (Philadelphia, American Philosophical Society, 1986), p. 88; S. F. Hockey, *Quarr Abbey and its Lands, 1132–1631* (Leicester, Leicester University Press, 1970), p. 40.

107 See the discussion in Thomas F. Glick, *Islamic and Christian Spain in the Early Middle Ages: Comparative Perspectives on Social and Cultural Formation* (Princeton, Princeton University Press, 1979), pp. 96–9.

108 See, for example, *Partidas* 3:32:18, where mills are licensed by the King and town councils, but mill owners must adhere to the obligation of all riparians not to impede the free flow of water.

economic interests of a lord to interfere with irrigation arrangements among his men, and sustained interference would cause such systems to break down. Lords unfamiliar with irrigation systems did in fact tamper with them when they acquired, from mid-thirteenth century on, fully-functioning Muslim-built systems. Those who attempted to privatise them or to monopolise the water soon enough became aware of the dysfunction they had induced and usually they corrected their mistakes. When they failed to, the systems could not function. The lords therefore learned how to maintain the autonomy of irrigation communities, along with their rules of equality and equity in the allocation of water, for their own benefit: the powerful who enjoyed 'disproportionate benefits from the institution of co-operation enforced the rules of the game and gave leadership to solidaristic efforts'.[109] This explains why magnates went to court on behalf of their irrigators, as Felip Boyl, lord of Massamagrell in the huerta of Valencia did, in a famous case in 1438 still recalled by Valencian irrigators.[110] Moreover, the local control requirement of irrigation is completely congenial to other aspects of 'feudalism', such as, for example, the typical fragmentation of jurisdictions which made it fairly easy to add another independent, cross-cutting jurisdiction, that of autonomous communities of irrigators.

The characteristic posture of a monastery can be appreciated from the case of San Pedro de Cardeña and its attempt to monopolise water use on Arlanzón in the tenth century. In mid-century, much of the water that Cardeña wanted was controlled by the monastery of San Martín del Río located in the village of Villavascones. In 945–950 twenty-four residents of the villa ceded their water rights, along with considerable land, to the monastery (a donation which may, in fact, have masked a purchase). These twenty-four appear to have been the original owners of an irrigation canal, or their successors, and were obliged to negotiate with the abbot in order to continue to receive water for irrigating their fields. The same abbot had also purchased another irrigation canal, dug by the parents of Diego of Ovécoz in Castrillo and then owned the channel from bank to bank (*de litus ab alio litus*); this canal had a mill called Micarri on it and represented a hydraulic 'package' of considerable worth. All this passed into the hand of Cardeña when it acquired the smaller monastery.[111]

109 Pranab Bardhan, 'Symposium on Management of Local Commons', *Journal of Economic Perspectives*, 7 (1993), 87–92, on p. 91.

110 Thomas F. Glick, *Irrigation and Society in Medieval Valencia* (Cambridge, MA, Harvard University Press, 1970), pp. 79–80.

111 Salustiano Moreta Velayos, *El monasterio de San Pedro de Cardeña: Historia de un dominio monástico castellano (902–1338)* (Salamanca, Universidad, 1971), pp. 47–8.

In Catalonia, there are many references from the ninth to the eleventh centuries of irrigated parcels called *insulae*, on the Llobregat (the river of Barcelona), the Ter (that of Girona), and the Segre (Seu d'Urgell). These 'islands' were fields in the beds of rivers, in places where the slope is very gradual and the river bed wide, particularly in the meanders of the lower courses of rivers or in their estuaries, which were either irrigated by flooding or by a network of surface canals. When not developed for irrigation they served as natural pastures. Many of these *insulae* were within the limits of cities and therefore owned by their bishops. But they had been developed by peasants, who had converted them from uncultivated *insulae* to *insulas edificatas* in the tenth century. Typically, fruit trees were grown there, as were legumes, grapevines, and cane. Some peasants lived on the premises. Martí demonstrates that cultivation of *insulae* was an ancient practice, recorded in the *Digest* and other ancient sources, and that the practice lasted because feudal lords were interested in developing communal irrigation systems.[112] In Al-Andalus, the place-name al-Jazīra (the island, as in the place-names Algeciras or Alcira, the latter a place highly developed for irrigation in both Muslim and Christian times) had a similar connotation, both as a site for irrigation agriculture as well as for mills.[113]

Even as Catalan village communities fell under feudal control, villagers were able to retain some measure of autonomy, particularly in matters of irrigation. Thus 'they nominated representatives ... who had sovereign power to arbitrate in conflicts relating to the employment of water and above all they appeared as the most dynamic element in both the economic management (especially hydraulic) of the countryside and in resistance to the feudal regime'.[114]

112 Ramon Martí, 'Les *insulae* medievals catalans', *Butlletí de la Societat d'Arqueologia Lul·liana*, 44 (1988), 11–23.

113 *Ibid.*, p. 121. See the Jazīra in the Paleoandalusí lower Segura: Rafael Azuar Ruiz, 'La rábita califal de Guadamar y el paleoambiente del Bajo Segura (Alicante) en el siglo X', *Boletín de Arqueología Medieval*, 5 (1991), 135–50, on p. 145; also on Algezira, on the banks of the Ebro, Miquel Barceló, 'Aigua i assentaments andalusins entre Xerta i Amposta (s. VII–XII)', *II CAME*, II, 413–20, on p. 415.

114 Bonnassie, *From Slavery to Feudalism*, pp. 257–8. For feudalism, the local control typically exercised by village irrigators was a two-edged sword. Self-governing irrigation systems were highly productive but encouraged values at odds with seigniorial control.

PART TWO

The late medieval and early modern cultural transitions

6

Reading the *Repartimientos*: modelling settlement in the wake of conquest

The *Libros de Repartimiento* and *Domesday Book*

In the wake of the conquest of Al-Andalus, the kings of Aragón and Castile set in motion a total reordering of the rural landscape. From the time of the conquest of Mallorca (1229) on, this reordering was documented in a distinctive series of land registers called the *Llibres del Repartiment/Libros de Repartimiento*, which are certainly unprecedented in any country of medieval Europe because they provide a moving picture over four centuries which document a cross-cultural transfer of landscape. In a certain sense, so does the *Domesday Book*, inasmuch as that famous inventory was supposed to establish the pattern of English land tenure as it had been in the time of the Anglo-Saxon King Edward. But *Domesday*'s rationale was quite different: it was first and foremost an instrument whereby the King might know what feudal dues were owed him. But both series share the registration of landholds across a cultural divide.[1] A prime distinction here, however, is that while *Domesday* records fiefs, many *Repartimientos* record the lowest level of landholding, down to individual peasant parcels (or to units – *cuadrillas* – of peasant parcels). This was because, unlike the situation in England where ceorls/villeins remained on their ancestral holds, the Castilian and Aragonese monarchs had to replace, in many places, one entire population by another. This is especially true of Andalucía, where

1 There has been some minimal interest by Domesday scholars in land registers of Norman Sicily, but only with regard to what in my view is a trivial point: they wanted to know if Normans elsewhere relied on pre-conquest land registers, whether Saxon or Arab. There are more interesting questions to be asked of Sicilian registers. See D. Clementi, 'Notes on Norman Sicilian Surveys', in V. H. Galbraith, *The Making of Domesday Book* (Oxford, Oxford University Press, 1961), pp. 55–8, and Sally Harvey, 'Domesday Book and its Predecessors', *English Historical Review*, 86 (1971), 753–73, on p. 765. Like the Spaniards, the Normans preserved the *antiquas divisiones Sarracenorum* when establishing their rule. See Henri Bresc, 'Féodalité coloniale en terre d'Islam: La Sicile (1070–1240)', in *Structures féodales et féodalisme dans l'Occident méditerranéean (Xe–XIIIe siècles): Bilan et perspectives de recherches* (Rome, Ecole Française de Rome, 1980), pp. 631–47, on p. 635.

the kings of Castile attempted to implant a free peasantry. Thus the usability of houses and estates by settlers was perhaps the overriding criterion of the *Repartimientos.*[2]

The *Domesday Book* records two very different principles of settlement and land distribution, one seigniorial, the other territorial.[3] While the distinction also appears in the *Repartimientos,* the emphasis is on territorial rather than seigniorial space, although there is some notational overlap, to be sure. We can also recognise that the Domesday series is generally richer in the kind of raw economic and social data that medievalists like and are familiar with. Much of this information is lacking, or only sporadically provided, in the *Repartimientos,* particularly in sixteenth-century registers for Andalucía where not only fruit trees, but in many instances native trees also, are enumerated. Counting trees picks up a tradition of the Roman census wherein, according to the *Digest,* the number of vines and olives trees were to be counted. Taking the *Repartimientos* as a whole, as representing a particular tradition of public administration, one gets the feeling that the entire enterprise was informed by Roman law tradition, as if the genre had been devised by Roman lawyers familiar with the criteria the Romans employed for drawing up tax lists. Thus Lactantius's complaint about the thoroughness of Roman *censitores,* who 'measured fields marker by marker, counted grapevines and trees, registered animals of all kinds, noted the names of men in the cities', is reminiscent of the modus operandi of the *apeadores,* or at least the most conscientious of them. The Roman practice of classifying land by categories is also reproduced in the *Repartimiento* tradition.[4]

Culturally, England remained substantially Saxon, whereas Spain underwent a profound cultural change. As a result, and in view of the general lack of detailed local records in Arabic, the

2 Julio González notes ruined vills or houses in New Castile: Julio González, *Repoblación de Castilla la Nueva,* 2 vols. (Madrid, Universidad Complutense, 1975–76), II, 284–8. See also Manuel Barrios Aguilera, *Libro de los Repartimientos de Loja* (Granada, Universidad, 1988), p. 59: '*alcaria caida*'. There is an analogy here with Domesday which notes whether a field or vill is devastated (*wasta est terra, mansiones vastatae,* etc.), but again the interest is in value, not usability *per se*; H. C. Darby, *Domesday England* (Cambridge, Cambridge University Press, 1977), pp. 234, 236, 238.

3 See Robin Fleming, *Kings and Lords in Conquest England* (Cambridge, Cambridge University Press, 1991), pp. 154–62.

4 See John Percival, 'The Precursors of Domesday: Roman and Carolingian Land Registers', in Peter Sawyer, ed., *Domesday Book: A Reassessment* (London, Edward Arnold, 1985), pp. 5–27, on pp. 12f. Citation from Lactantius in Josep M. Salrach, 'Del estado romano a los reinos germánicos. En torno a los bases materiales del poder del estado en la Antigüedad tardia y la alta edad media', in *De la Antigüedad al Medioevo* (Madrid, Fundación Sanchez-Albornoz, 1993), pp. 95–142, on p. 103.

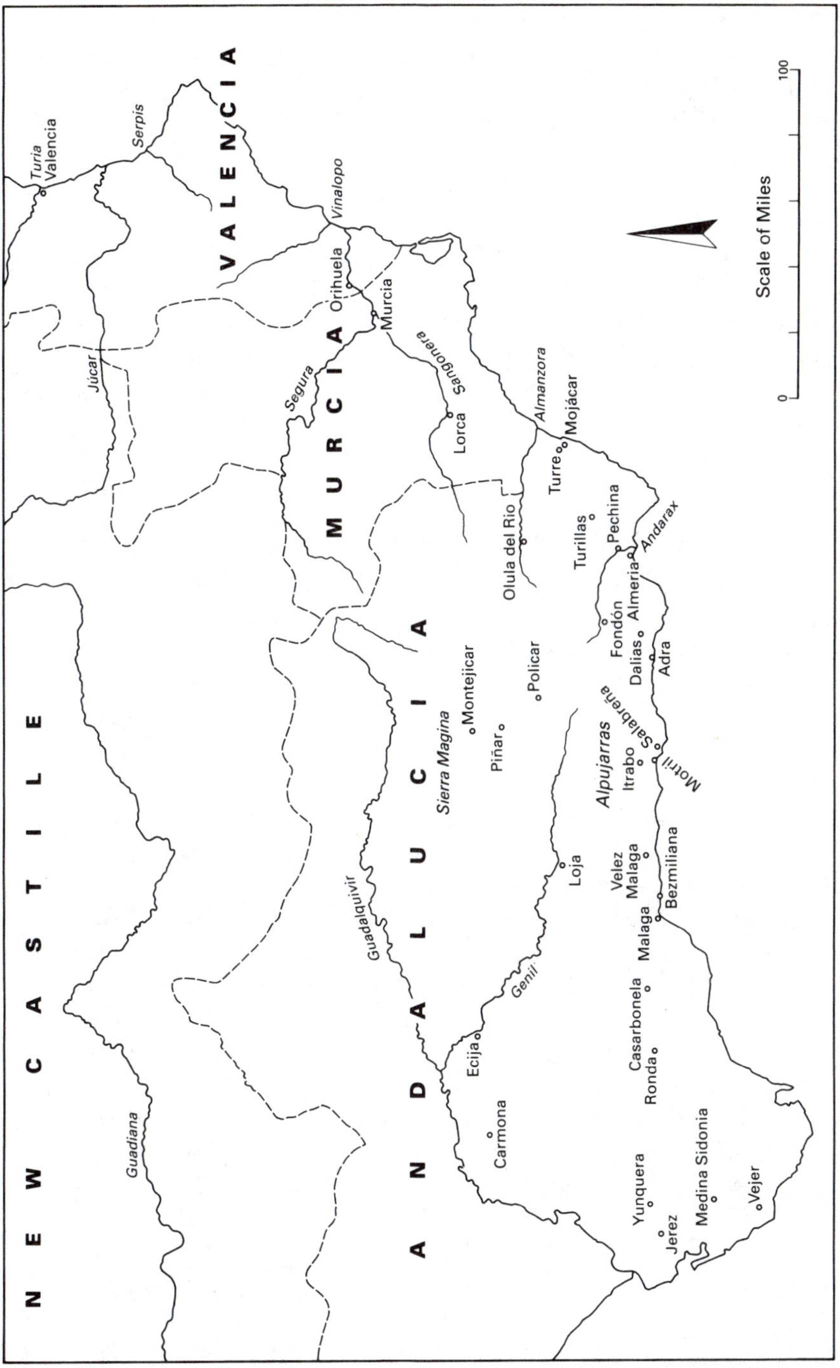

Map 8 *Repartimientos*

Repartimientos are invaluable both as a source for recreating the lost landscape of Al-Andalus and for evaluating the ensuing modal change in the ordering of the countryside.

In the following discussion, I will show to what extent the *Libros de Repartimiento* document an Andalusi model of landholding, and in what ways they reveal (or do not reveal) the replacement model that the Christian conquerors imposed on the conquered landscape. In this endeavour, I will not only examine the medieval registers, but will carry the story forward through the sixteenth-century books of *Apeo* and *Repartimiento* of Granada in which were recorded the land transfers taking place after the expulsion of the Moriscos from the Kingdom and which, both technically and conceptually, represented a continuation of the same process of landscape change already described.

Repartimiento before the *Libros*

I will begin my review of the *Repartimiento* literature with a brief discussion of the partition of property in Toledo following the conquest of that city, a process which was accomplished without the compilation of such registers or which did not require them to be preserved, even if some must have been generated at the time.[5] The area south of Toledo, as described by Julio González, consisted in large part of *alquerías* which had been abandoned for so long that the names of the former owners were not known and the structure of property was difficult to determine. *Repartimiento* was accomplished by Juntas de Partidores in Toledo and in the towns that adopted its Fuero. These Juntas held inquests and dispatched surveyors (called *sexmeros, quinoneros, cuadrilleros,* etc.) whose job was to break down larger units into smaller ones.[6] González notes that in large *secano,* wheat-farming *alquerías,* the fields were undivided or with only a few divisions. If such *alquerías* were donated whole, they were divided into halves, thirds or fourths, with further subdivision left up to the grantee. There was also a tendency in the time of Alfonso VII to join two *alquerías* together, to make villages of a certain size and density, a process which, according to González,

5 There is also a technological reason for the lack of any *Libro de Repartimiento* and why the *Repartiment* of Valencia (1238) was the first such register: the 'paper revolution' that had such a profound effect on royal administration began only after the Muslim paper mills at Játiva had fallen into King Jaume's hands. See Robert I. Burns, *Society and Documentation in Medieval Valencia* (Princeton, Princeton University Press, 1985), pp. 48–50, 151–61. The Repartiment of Mallorca (1232), like the eleventh-century Domesday Book, is written on vellum.

6 González, *Repoblación de Castilla la Nueva,* II, 162, 164, 174, 175, 180f.

broke the structure of quite a few latifundios encountered by the Castilians.[7]

What González appears to have been describing here, but was unaware of, was territory which had no metes and bounds, not because it was comprised of latifundia but which rather consisted of *alquerías* that had been held previously by clan groups farming collectively. It is probable that even in this dry-farming region, the countryside had been organised in *ḥiṣn/qarya* complexes. González documents a number of place-names in *ḥiṣn*, such as Exnavexore, or Aznaron (=*ḥiṣn* Hārūn), as well as the descriptive use of the word: the citadel of Toledo was called Alhicen; houses were donated in Monzón, *in illo alhizen de illo castello* (1090) and in Calahorra (1074), within the Alhicen.[8] Unlike the situation in eastern Spain, however, the Muslim *ḥuṣūn* did not become nodal points in the resettlement of this region and most fell into disuse.[9]

González's account of the rural landscape of New Castile is similar to his earlier account of the *Repartimiento* of Seville, which also suffered from the lack of a model of agrarian space in Al-Andalus. Here, in spite of the massive documentation afforded by that *Repartimiento*, González's description of the rural landscape is strangely disjointed. There is no sense of a hierarchy of settlements. His definition of *alquería* here is 'A rural entity which maintained unity, at least partial, of property. The majority were preserved whole.' The concept of unity is basic for appreciating not only the physical deployment of the settlement (boundaries, parcellisation patterns), but also to the built areas.[10] This unity of structure, however, he associated with Roman centuriation (the prevalence of lots of 30 yugadas suggested a regularity of partition within the logic of centuriation).

It is more to the point, rather than bracketing Andalusi history and presuming such scarcely unchanged Roman surveying patterns, to associate the features he describes – once again – with *alquerías* which, in origin at least, had been undivided. Certain *alquerías* or parts thereof did indeed have Beni- names (e.g., *barrio* of Benimahmut).[11]

7 *Ibid.*, II, 176, 178, 311. The amalgamation of smallholdings into more viable units and 'the combining of divided vills into a single holding' were hallmarks of the Domesday grants (Fleming, *Kings and Lords*, pp. 122, 150).

8 *Ibid.*, pp. 224, 291. Houses *within* the *ḥiṣn* suggest that an *albacar* is referred to. Compare the identical wording from an Aragonese document of 1102: 'casas quas te feceris in illo alluzem (al-ḥiṣn) de illo castello'; Carlos Laliena and Philippe Sénac, *Musulmans et Chrétiens dans le Haut Moyen Age: Aux origines de la reconquête aragonaise* (Toulouse, Minerve, 1991), p. 61 n. 61.

9 Jean-Pierre Molénat, 'Villes et forteresses musulmanes de la région tolédane disparues après l'occupation chrétienne', *Castrum 3*, pp. 215–24.

10 Julio González, *Repartimiento de Sevilla*, 2 vols. (Madrid, CSIC, 1951), I, 396.

11 *Ibid.*, III, 416; II, 44.

There are also a number of suggestive *ḥiṣn* place-names in the jurisdictional district (*alfoz*) of Seville, including Aznalfarache, Aznarcóllar, and Aznalcázar.[12] What the *Repartimiento* of Seville describes is an Islamic landscape, deprived of its social basis, in the process of losing its coherence, as *alquerías* were merged, absorbed into the city, and so forth. González also describes, without fully understanding their significance, parcels called *machar* (Ar. *majshar*), which were compounded with personal names, such as Machar Almanzor or Machar Alcadi (*machar* of the judge).[13] These estates or *cortijos* would seem to be the equivalent of the Valencian and Murcian *raḥal*: a single estate owned by a wealthy or prestigious individual, presumably associated with the State.

González, in both the cases of New Castile and of Seville, gives an account of the pre-existing landscape which is too undifferentiated to provide any conceptual basis for analysing the change in model. He realises there must have been some underlying organisational model and chooses, inappropriately in my view, a Roman one.

The *Repartiment* of Mallorca

The baseline for the evaluation of the Andalusi rural landscape is established in the earliest of the books of *Repartimiento*, that of Mallorca, which survives in Arabic, Latin, and Catalan versions.[14] In

12 *Ibid.*, I, 388. Note the curious semantic doubling in Aznalcázar (= *ḥiṣn al-qaṣr*).

13 *Ibid.*, I, 423; II, 45. Cf. a *Raḥal al-qāḍi* in Valencia (Pierre Guichard, *Les musulmans de Valence et la Reconquête (XIe–XIIIe siècles)*, 2 vols. (Damascus, Institut Français de Damas, 1990–91, II, 384).

14 For the Arabic version, see Jaime Busquets Mulet, 'El códice latinoarábigo del Repartimiento de Mallorca (texto árabe)', in *Homenaje a Millás Vallicrosa*, 2 vols. (Barcelona, CSIC, 1954–56), I, 243–300. Busquets also published the Latin text: 'El códice latinoárabigo del Repartimiento de Mallorca (parte latina)', *Butlletí de la Societat Arqueològica Lul·liana*, 30 (1947–52), 6–55. For the Catalan version, Ricard Soto i Company, ed., *Còdex català del Llibre del Repartiment de Mallorca* (Palma de Mallorca, 1984). Although the conquerors must have had access to Arabic land *dīwān/s* in some places, this is the only surviving Arabic specimen, although it covers only Palma and its immediate environs. All three versions have the limitation of covering only royal land. Hence, a good complementary document is the register of lands granted to an important magnate, Antoni Mut Calafell and Guillem Rosselló Bordoy, eds., *La remembrança de Nunyo Sanç: Una relació de les seves propietats a la ruralia de Mallorca* (Palma de Mallorca, Museu de Mallorca, 1993). M. Barceló ('Vísperas de feudales. La sociedad de Sharq al-Andalus justo antes de la conquista catalana', in Felipe Maíllo Salgado, ed., *España, Al-Andalus, Sefarad: Síntesis y nuevas perspectivas* (Salamanca, Universidad, 1988), pp. 98–112, on p. 101) observes that the surveyors did not understand Muslim social space and frequently did not measure what they regarded as interstitial spaces.

the *Repartiment*, all the measures are given in jovates (1 = 11.36 ha), so it is possible to derive ratios among different kinds of settlements. Poveda Sánchez compared *alquerías* to *raḥal/*s and found that the former were considerably larger (7 versus 4 jovates, or 83.7 v. 49 ha). The distinction between *alquería* and *raḥal* was unclear to the Catalans and almost immediately they lost their specific meanings.[15] He then looked at *alquerías* with Beni- names (92 of them from the *Repartiment*, plus 98 more from complementary documentation; 65 *raḥal/*s also had Beni- names), and found that in area they clustered around the mean for all *alquerías* of 7.63 jovates. That means that statistically, Beni-named *alquerías* represent the mean type of exploitation.[16] The proper names compounded with Beni- suggest that the majority of these settlements were Berber due to the late occupation of the island and known waves of Berber settlement under the Almoravids and Almohads.[17]

The Christians changed the agrarian landscape immediately. In Mallorca, wherever there was a preponderance of irrigation agriculture, there is a general presumption of a retrocession of irrigation, not because Christian peasants didn't know how to irrigate or weren't as adept at it (as the myth goes), but because feudal rent, which was normally taken in measures of grain, was the tail wagging the dog of peasant settlement. The feudal tax structure, that is, demanded a certain level of investment in cereal culture. Whether such crops were irrigated or not (as they certainly were in Aragón) is another question.[18] The demand for grain also sets up a situation of positive feedback where the need for additional water for milling acts as a further constraint on irrigation. Nor was there any land in Muslim Mallorca specifically set aside for grape-vines, which were considered just another garden crop. The lords did not like garden vegetables, which were difficult to commercialise. Hence they did not tax village *horts*, a further indication of how changing tastes and values can force a severe shift in agrarian regimes. Land-use in thirteenth century, post-*Repartiment* Mallorca was approximately 31 per cent in irrigation, 30 per cent in vineyards, 30 per cent in arboriculture, and an unknown amount in dry-farmed cereals. We can deduce, since Christians for all intents introduced grapes,

15 Angel Poveda Sánchez, 'Introducción al estudio de la toponimia árabe-musulmana de Mayurqa según la documentación de los Archivos de la Ciutat de Mallorca (1232–1278)', *Awraq*, 3 (1980), 75–102, on p. 96.

16 *Ibid.*, pp. 84–5, 95–6.

17 *Ibid.*, pp. 83, 95.

18 See, in this regard, Pierre Ponsot, 'Les Morisques, la culture irrigué du blé et le problème de la décadence de l'agriculture espagnole au XVIIe siècle', *Mélanges de la Casa de Velázquez*, 7 (1971), 237–62.

that Muslim agriculture had been split between irrigated *horts* (on which some cereal grains were grown) and arboriculture.[19]

In terms of the morphology of rural settlement, grants were in the form of dispersed fragments; the process of *repartimiento* tended towards parcellisation of *alquerías* presumably encountered in the form of undivided, communally worked fields.[20]

The *Repartiment* of Valencia

The Valencian *Repartiment* covers the entire kingdom.[21] In Valencia both *alquerías* and *raḥal/*s are well defined. *Alquerías* typically had ten to fifty houses: in irrigated areas they were quite small (0.5 to 2.5 km; while some of the mountainous tree-growing *alquerías* were quite large, 9 km square (e.g., Benilloba, 9.26 km; Benasuau, 9.5 km). Of 150 agricultural places in the huerta of Valencia mentioned in the *Repartiment*, two-thirds are *alquerías*, one-third *raḥal/*s. The latter were much smaller than the former and had personal names.[22] *Alquerías* were collectively worked and had no fixed territorial limits. Christian settlers were normally granted three jovates (*c.* 9 ha), which were agriculturally mixed, including some huerta and vineyards near the settlement, cereal fields farther away. Whenever it was the rule to mix land use in the composition of lots, that implied in practice the maintenance of parcellary dispersion and of the size of pre-existing parcels.[23] Therefore, the Christian post-*Repartimient* agrarian landscape tended to physically resemble the Muslim one preceding it. But land tenure was quite a different matter. Directly after the *Repartiment* immediate social stratification ensued under conditions of a very active market in land. The atomisation of parcels that followed made the Islamic *alquería* unrecognisable, as individual ownership replaced collectivity as the organising principle of tenure. As had been the case in Mallorca, Christian cultivators concentrated their efforts on cereals and grape-vines, to the prejudice of huertas (even though these were mainly free of

19 Ricard Soto i Company, 'Repartiment i Repartiments: l'ordenació d'un espai de colonització feudal a la Mallorca del segle XIII', in *De Al-Andalus a la sociedad feudal: Los repartimientos bajomedievales* (Barcelona, CSIC, 1990), pp. 1–51, on pp. 37–8.

20 *Ibid.*, p. 30.

21 María Desamparados Cabanes Pecourt and Ramón Ferrer Navarro, eds., *Libre del Repartiment del Regne de Valencia*, 3 vols. (Zaragoza, Textos Medievales, 1979).

22 Pierre Guichard, in *Nuestra historia* (Valencia, 1980), II, 269f., and *Musulmans de Valence*, II, 375–85.

23 Manuel Barrios Aguilera and Margarita M. Birriel Salcedo, *La repoblación de Granada después de la expulsión de los Moriscos* (Granada, Universidad, 1986), p. 52. I believe this rule holds true for medieval, as well as for early modern *Repartimientos*.

dues). It was at this point, in Torró's view, that Mudéjars, in order to pay their taxes, began irrigating wheat. In that wheat could be dry-farmed, some Muslims petitioned to have their fields changed from irrigated to unirrigated status.[24]

In a masterful study of the feudal transformation in Alcoi, Torró shows how this new town, of Christian foundation, rose on the same space formerly occupied by a *ḥiṣn/qarya* complex. Of twelve newly delimited agricultural zones surrounding the town, six corresponded to old Andalusi *alquerías*, occupying from between 72–90 ha, quite close to the mean of 83.7 that Poveda Sánchez had found in Mallorca. These settlements were irrigated by the Acequia de Barxell, whose course wound around under the *ḥiṣn*, now known as El Castellar. Forty per cent of the parcels granted out to Christian settlers were small, irrigated ones, half of which were in the Horta d'Alcoi. Grapes were planted on terraces that were lower with respect to huertas and irrigated with excess water. There was a vast expansion in cereal cultivation over what the Muslims had formerly cultivated. As elsewhere, there was an immediate tendency towards fragmentation of holdings and dispersion of parcels.[25]

With respect to *alquerías*, there were two modes of partition. Either they were seigniorialised, granted whole to lords, or they were apportioned by small parcels. The latter mode was much more disturbing to Mudéjar society. Even when, directly after the conquest, Muslims remained in place, their *alquerías* were atomised into single family units: the original undivided and unbounded *alquerías* were now surveyed and parcelled out.[26] Thus, although the *Repartiment* may give a sense of continuity of population in Mudéjar rural communities, discontinuity was apparent: the parcellisation of *alquerías* represents a formal pressure towards social destabilisation (detribalisation).

In the *Repartiment* of Valencia, antecessors were generally mentioned.[27] The reasons for specifying Muslim antecessors were both practical and jurisprudential. Pre-existing units could be identified without having to be resurveyed. Thus 'a field belonging to

24 Josep Torró, *Poblament i espai rural: Transformacions históriques* (Valencia, Diputació Provincial, 1990), pp. 81–6, 95–6.

25 Josep Torró, *Alcoi. La formació d'un espai feudal (de 1245 a 1305)* (Valencia, Diputació Provincial, 1992), pp. 45, 50, 162, 208, 211–12, 265.

26 Pierre Guichard, 'La conquista militar y la estructuración política del reino', *Nuestra historia*, III, 29; Josep Torró, 'Sobre ordenament feudal del territori i trasbalsaments del poblament mudèjar. La *Montanea Valencie* (1286–1291)', *Afers*, 7 (1988–89), 95–124, on p. 107.

27 Antecessors also appear regularly in the *Repartimientos* of Murcia, 'Almería, Comares, Turillas, and Yunquera, among others.

Hamete' was a kind of convenient shorthand specifying the parcel's exact bounds. Legally the rationale was more complex. First, continuity of tenure units and the orderly takeover that this implies constituted a legitimisation of conquest. Second, the invocation of an antecessor, together with the royal grant that accompanied *repartimiento,* was the basis of legal title. Third, the antecessors collectively were the frame of reference for the establishment of title to land and of both the validity and priority of water rights: in this sense tracing a title to 'the time of the Moors' was the Spanish equivalent of the English Norman standard of 'the day on which King Edward lived and died'. As Sir Frank Stenton noted, 'the written administration needed for such a transfer [between antecessor and successor] is placed at a minimum, and the responsibility for determining landed rewards is neatly placed on the grantees rather than on the representatives of royal government'.[28]

Orihuela and Murcia

These were *Repartimientos* ordered by Alfonso X, even though Orihuela soon after became part of the Crown of Aragón. In Orihuela there was a wholesale replacement of the Muslim population by Christians. There was a general policy of aggregating two, three and up to six *alquerías* to make a *cuadrilla,* which became the basis for a new administrative unit. An interesting aspect of the *Repartimiento* of Orihuela is that the royal officials found substantial tracts of land which had 'never been surveyed in the time of the Moors'. All of this land lay in *alquerías* with Beni- names: Benmira, Benamoquetib, Benijuües, and so forth. One settler is recorded as wanting to fence (*tapiar*) his property, another indication of an *alquería* that had previously been undivided. The fields in question were in the limits between the huerta and the sea, marshy land that is difficult to stabilise for irrigation.[29]

In Murcia, where the *Repartimiento* only covers the huerta itself, there was a high incidence of minifundios (that is, parcels of no more than 200 sq m), reflecting the high level of parcellisation typically prevailing in highly productive irrigated areas of

28 Quoted by Fleming, *Kings and Lords in Conquest England,* p. 110.

29 Juan Torres Fontes, 'Los Repartimientos murcianos del siglo XIII', in *De Al-Andalus a la sociedad feudal: Los repartimientos bajomedievales* (Barcelona, CSIC, 1990), pp. 71–94, on p. 88; Juan Torres Fontes, ed., *Repartimiento de Orihuela* (Murcia, Academia Alfonso X el Sabio, 1988), p. 89 ('Fue fincada por mala terra que negun la quiso tomar en las otras particiones et auya y dellas muchia que en tempo de moros nonqua foron sogueadas').

Al-Andalus.[30] Here again, the traditional *alquería* array was broken by the composition of *cuadrillas* as the basis of a new administrative order. Here we detect something like what F. W. Maitland, in reference to *Domesday*, called 'notational movability' of parcels.[31] That is, some *cuadrillas* never acquired judicial independence, but were integrated into neighbouring ones. These *cuadrillas* were organised to fit the irrigation system; each one was dependent on a specific acequia which acquired the name of the parcels it irrigated.

Andalucía[32]

To return to the *Repartimiento* of Seville: it is mainly a rural document which shows (according to González Jiménez) the concentration of property in a few hands in Almohad times.[33] I must again express my doubts about the prevalence of Muslim latifundia and wonder whether these same documents can be read for evidence of undivided *alquerías*, which is just what one would expect of recently settled Berber tribesmen, as in Mallorca, for example.[34] The same author's account of the modal change in agrarian settlement would seem to support the latter view: with so much Muslim emigration in the wake of conquest, the old settlement pattern of rural population based on many small population nuclei – *alquerías*, that is – disappeared and was replaced by a much more concentrated pattern. Of 160 *alquerías* recorded in the Aljarafe, for example, only thirty were repopulated. Except for grants to a small number of major magnates (*donadíos*), which in any case were distributed through a number of villages, parcel size was small or middling.

30 Torres Fontes, 'Repartimientos murcianos', p. 76. The text of the *Repartimiento* was edited by Juan Torres Fontes, *Repartimiento de Murcia* (Madrid, CSIC, 1960).

31 F. W. Maitland, *Domesday Book and Beyond* (London, Collins, 1961), p. 33.

32 For overviews of medieval Andalusian *Repartimientos*, see Cristina Segura Graiño, 'Los Repartimientos medievales andaluces: Estado de la cuestión', *Anuario de Estudios Medievales*, 12 (1982), 625–39; and Manuel González Jiménez, 'Repartimientos andaluces del siglo XIII: Perspectiva de conjunto y problemas', in *De Al-Andalus a la sociedad feudal: Los repartimientos bajomedievales* (Barcelona, CSIC, 1990), pp. 95–117.

33 González Jiménez, 'Repartimientos andaluces del siglo XIII', p. 102.

34 That is not to say that latifundios were inexistent in Al-Andalus. José E. López de Coca Castañer (*La tierra de Málaga a fines del siglo XV* (Granada, Universidad, 1977), p. 44) gives examples of large property (e.g., estates of 534 and 715 ha in Bracalema and Comares, respectively) and of a *realidad latifundista* according to the modern criterion of any property more than 250 ha. One would have to make a careful study, region by region, and in different periods to test that reality. Meanwhile one must beware of reading into the past the present presumption of latifundios in Andalucía.

The cheapness of land and a booming land market created by *repartimiento* made possible the concentration of parcels, an incremental process which by the end of the fifteenth century resulted in the seigniorialisation of half the land in Andalucía. But clearly the *Repartimientos* themselves did not create latifundia.[35]

There are only a few places in the *Repartimiento* of Seville where metes and bounds of village bounds are recorded, supporting the supposition that *alquerías* tended not to be demarcated in Al-Andalus through the thirteenth century.[36] This can be appreciated in a characteristic document, that of the demarcation (*deslinde*) between Medina Sidonia, Jerez, Vejer, Tarifa, Algeciras, and Alcalá de los Gazules in 1269. Of these towns, Vejer, Medina Sidonia, and Alcalá seem clearly to have been *ḥuṣūn* with dependent *alquerías*.[37] The town bounds, which in any case were larger than had been the case under Islamic rule when common grazing lands between settlements were not demarcated, became extremely important under the Christians and had to be scrupulously measured because of the high value placed on grazing rights.[38] A *faqīh*, asked to comment on aspects of landownership and municipal jurisdiction before the conquest, said that everyone knew where the grazing areas were

35 González Jiménez, 'Repartimientos andaluces del siglo XIII', pp. 106–7, 113; Emilio Cabrera, 'Evolución de las estructuras agrarias en Andalucía a raiz de su reconquista y repoblación', in *Andalucía entre Oriente y Occidente (1236–1492)* (Córdoba, Diputación Provincial, 1988), pp. 178–89; Mercedes Borrero Fernández, 'Las transformaciones de la estructura de la propiedad de la tierra en la Baja Andalucía en la segunda mitad del siglo XIII', in *ibid.*, pp. 191–208. In Cabrera's account, latifundios begin at 300 ha.

36 Thus the heredamiento of Alcalá de Guadayra is one of the few places to be described with metes and bounds, suggesting it had been surveyed in Muslim times; González, *Repartimiento de Sevilla*, II, 128. See, on p. 359, a grant of Alfonso X specifying that the *término* of this place will be the same as it had been in the time of the Moors.

37 Miguel Angel Ladero Quesada and Manuel González Jiménez, 'La población en la frontera de Gibraltar y el Repartimiento de Vejer (siglos XIII y XIV)', *Historia, Instituciones, Documentos*, 4 (1977), 199–317, on p. 266, ask whether *aldeas* represented the structuring of the rural population of this area before the conquest or whether the King ordered the *partidores* to organise the territory around a fortified nucleus and a series of small settlements dispersed throughout the administrative limits of the town, as had been done earlier in Ecija. It seems to me more likely that in the three cases mentioned at least, a pre-existing *ḥiṣn/qarya* complex was the operative model. On Ecija, see María Josefa Sanz Fuentes, 'Repartimiento de Ecija', *Historia, Instituciones, Documentos*, 3 (1976), 533–51: the término was set at one league surrounding the town; beyond this area was a broader district, including the *aldeas*, which had been marked off in Islamic times. The *partidor*, accompanied by sons of the former Muslim *alcaide*, placed markers for the villages where they had been 'in the time of the Moors, their fathers and grandfathers'.

38 Ladero Quesada and González Jiménez, 'Repartimiento de Vejer', pp. 226–30.

and these were not controlled by individual towns, 'because among the Moors there were no districts marked off for this purpose'.[39]

Between the *Repartimiento* of Seville and that of Loja (1486), where metes and bounds are provided for many places, it is clear that the nature of land tenure in Al-Andalus had been changing, evolving towards individual ownership and juridical definition of *alquerías* and towns, associated no doubt with the breakdown of tribal society and its typical modes of land tenure and village organisation. By the end of the fifteenth century, *alquerías* (in Nasrid Granada) had lost their homogeneous and communal nature, endogamy yielded increasingly to exogamy, and as a result *alquería* parcels were increasingly dispersed, a trend enhanced by the ever-growing roster of parcels held by *habices* or religious trusts. The following elements of the earlier village structure remained: first, its general topographic layout, morphology and spatial organisation; second, the basic autonomy between urban and rural worlds, with scant penetration of the *alquería* by urban groups and scant dependence of *alquerías* on cities; third, differentiation of territories of *alquerías*, each with its own bounds, although there were shared spaces among them, notably those for pasturage.[40]

When Christian *partidores* surveyed Loja in 1487 along with a half a dozen Muslim former residents, sent by the King of Granada, it is clear that boundary markers between villages were already in place, as the Muslims guided the Christians from one to another. Nasrid village bounds were observed if the Muslim population remained, not necessarily when the district had been evacuated.[41] In both Loja and Comares town boundaries were determined not only by obvious markers like watch towers (*atalayas*), small forts and wells, but by the broad contours of watersheds, with water flowing in one direction marking the bounds of one town, that in the opposite direction, its neighbour.[42] It is also clear, in the case of Comares, that grazing land – still communal – had never been surveyed, inasmuch as the Christian officials are depicted placing

39 Manuel Acién Almansa, *Ronda y su serrania en tiempos de los Reyes Católicos*, 3 vols. (Málaga, Universidad, 1979), II, 610.

40 Antonio Malpica Cuello, 'De la Granada nazarí al reino de Granada', in *De Al-Andalus a la sociedad feudal* (Barcelona, CISC, 1990), pp. 119–53, on pp. 122, 133, and 'Poblamiento del Reino de Granada. Estructuras nazaríes y modificaciones castellanas', in *Les Illes Orientals d'Al-Andalus* (Palma de Mallorca, Institut d'Estudis Baleàrics, 1987), pp. 375–93, on p. 201.

41 López de Coca Castaner, *Tierra de Málaga*, p. 124.

42 Barrios Aguilera, *Libro de los Repartimientos de Loja* (n. 2, above), pp. 56, 59. Francisco Bejarano-Robles and Joaquín Vallvé, *Repartimiento de Comares (1487–1496)* (Barcelona, Universidad, 1974), pp. 2–4, 42f.

markers along sheep runs (*cañadas*) and measuring them with cords (*sogas*).[43]

The fifteenth-century Granadan *Repartimientos* reveal an increasing pace of privatisation of agricultural property resulting in a modal change in the organisation of *alquerías* in the Nasrid kingdom. The other two elements of the *ḥiṣn/qarya* complex – the castle and the irrigation systems – had also changed. First, more castles may have been permanently garrisoned because of defence considerations. Second, the privatisation of water rights documented in the *Libros de Habices* suggests substantial degradation in the Berber communal model of rights, although that did not influence the operating procedures of irrigation systems.

Questions of tenure aside, the Christian conquest of Granada appears not to have induced much change in the Muslim agricultural system. With regard to the eastern mountains (Yznalloz, Piñar, and Montexicar), Peinado finds no evidence that Castilian settlers eliminated fruit trees, nor did they consider irrigated fields to be cereal space. The Nasrids themselves had encouraged the irrigation of cereals in order to reduce chronic flour deficits. Thus the agricultural style later attributed to the Castilians already existed in this area before they arrived.[44]

In the two centuries that intervened between Ibn al-Khaṭīb's description of Granada and the early sixteenth century, the number of *alquerías* in the vega diminished from 300 to 71.[45] After the Christian conquest, Morisco *alquerías* continued to be characterised by the predominance of clans, by the cultivation of fields according to traditional techniques, and the communal use of pastures. However, the clan was not the strong tribal segment of old and family groups tended to disperse and to own property in more than one *alquería.* There was an enormous admixture of lineages throughout fifteenth-century Granada, to the point where the correspondence between the founders of a place and its current population had practically disappeared, even though Luna Diaz found some *alquerías* in which one family owned between a quarter and half of the land. Once undivided family property was fragmented by inheritance,

43 Bejarano-Robles and Vallvé, *Repartimiento de Comares*, pp. 64ff.

44 On irrigation of cereals, Rafael G. Peinado Santaella, *La repoblación de la tierra de Granada: Los montes orientales (1485–1525)* (Granada, Universidad, 1989), pp. 103–4. See also Antonio Malpica Cuello, *Turillas, alquería del Alfoz seixitano* (Granada, Universidad, 1984), p. 26, and Rachel Arié, *L'Espagne musulmane au temps des Nasrides (1232–1492)* (Paris, Brocard, 1973), p. 346 (irrigation of wheat in Orgiva, Ferreira, and Poqueira – all in the Alpujarras).

45 Juan Andrés Luna Diaz, 'La alquería: un modelo socio-económico en la vega de Granada. Aproximación a su estudio', *Chronica Nova*, 16 (1988), 79–100, on p. 81.

dowry, or sale. Social status came to rest increasingly on wealth rather than on lineage.[46] From fifteenth- as well as sixteenth-century evidence therefore, it is clear that as tribalism broke down, privatised landholdings emerged, as demonstrated by the availability of private properties for religious trusts.

Luna Diaz found the mean number of houses in Granadan *alquerías* of the sixteenth century to be 140, with a density of 125 persons per square kilometre in irrigated areas, 13 for unirrigated. He found tremendous parcellisation of agricultural land throughout the vega: in *alquerías* near the city the average parcel size was 2.9 sq m, with average holdings of around 0.82 ha per proprietor.[47]

The resettlement of Granada after the expulsion of the Moriscos

After the second Alpujarras revolt, the Moriscos were expelled from the Kingdom of Granada and all their possessions passed into royal ownership by a provision of 24 February 1571. A supplementary royal order, issued on 27 September, stipulated that previously existing agricultural practices were to be preserved: new settlers had to farm according to the custom of their new place of residence and to preserve irrigation arrangements intact.[48] The subsequent partition process left something like a photographic record of Morisco agriculture which, in more than a few particulars, was the descendent of medieval settlement patterns and agricultural practice. The difference between the *Repartimientos* of the 1490s and the vastly more detailed systematic records of the 1570s is explained by the development of bureaucratic practices in the intervening years. The reassignment of property following the expulsions of the 1570s was broken down into two discrete processes, first the *apeo*, then the *repartimiento* itself. The *apeo*, as its etymology suggests, was something like the medieval field perambulation, a procedure by which metes and bounds of discrete properties were recorded, and a general inventory along the lines of *Domesday Book*. Those who carried out the inventory were not necessarily the same officials who divided the real estate. The *apeo* was executed immediately after the expulsion; there were preparatory dispositions, then a survey by administrative districts (*pagos*), and then notification was sent to

46 *Ibid.*, pp. 92, 95; Malpica Cuello, *Turillas*, pp. 15–16, 22–3; Manuel Acién Almansa, 'Reino de Granada', in Miquel Barceló, ed. *Historia de los pueblos de España. Tierras Fronterizas, I. Andalucía, Canarias* (Barcelona, Argos Vergara, 1984), p. 49.
47 Luna Diaz, 'La alquería', pp. 89, 92.
48 Barrios Aguilera and Burriel Salcedo, *Repoblación de Granada*, pp. 32, 39.

the Old Christians living in the area already. The *Repartimiento* took place some years later, with a carefully conceived arrangement of lots, sometimes with a master plan drawn up by sectors. Mosque property – *habices* – now in the hands of the Church was carefully noted and integrated into the master plan; provisions were made for a general census of population as well as for assignment and collection of taxes, and a procedure was laid out for follow-up inspections and emendations of the original grants. Because of the Castilians' interest in sheep-herding, *apeos* gave very precise descriptions of common land, and the *repartimientos* were 'generally more systematic [than the *apeos*] although less descriptive'.[49]

Although in the *Libros de apeo, alquerías* are called *lugares,* there was no substantive change either in the morphology of villages or in their organisation.[50]

Certain characteristic parcellisation patterns associated with the Morisco *minifundio* are described in *apeos.* The custom of distributing irrigated lands in different administrative districts (*pagos*) responded to the older Muslim custom of an individual's owning land in different *pagos,* to compensate for different qualities of land, and acted both to stimulate the fragmentation of parcels and to preserve the Morisco *minifundio.*[51] Some parcels belonged, collectively, to several Moriscos, no doubt members of the same family who, because of the smallness of their inheritance, preferred to maintain the land undivided.[52] Some small parcels bordered others of the same owner, who did not concentrate them. Cabrillana believes that, given the economics of Morisco landholding, small lots could be sold more easily.[53] The dispersion of parcels was so pronounced in some areas that some Moriscos owned trees on other's people's land, and single trees could have more than one owner.[54] In the *alquería* of Benamocarra (Velez Málaga) in 1486, there were

49 Manuel Barrios Aguilera, 'La repoblación del Reino de Granada por Felipe II', in Miguel Angel Laderro Quesada, ed., *La incorporación de Granada a la Corona de Castilla* (Granada, Diputación Provincial, 1993), pp. 607–28, on pp. 617–21.

50 Manuel Barrios Aguilera, *Moriscos y repoblación en las postrimeras de la Granada islámica* (Granada, Diputación, 1993), p. 208.

51 Pedro Ponce Molina, *El espacio agrario de Fondón en el siglo XVI* (Fondón, Ayuntamiento, 1984), p. 25. Repartidores, since the fifteenth century, had been enjoined to reckon 'bueno por bueno y malo por malo' when apportioning fields, so that all settlers shared in better and poorer grades of land (López de Coca Castañer, *Tierra de Málaga,* p. 295).

52 Nicolás Cabrillana Cizar, *Moriscos y cristianos en Yunquera* (Málaga, n.p., 1994), p. 53: 'un tablero que fue de Bartolomé Izmael e de Pedro Romero (= al-Ḥājj ?) e de Bernardino Zuleyman . . . todos juntos'.

53 Cabrillana Cizar, *Yunquera,* p. 54.

54 José Domingo Lentisco Puche, *La repoblación de Olula del Río (Almería) en el siglo XVI* (Almería, Instituto de Estudios Almerienses, 1991), p. 71.

Table 2 *Morisco and Christian settlement in sixteenth-century Granada (by parcel size)*

Area	*Moriscos*		*Settlers*	
	No.	%	*No.*	%
0–1	5	10	0	0
1–5	26	51	0	0
5–10	10	20	0	0
10–15	2	6	31	86
15–30	7	12	4	11
30+	1	1	1	3

parcels consisting of one olive or one almond tree only. The sixty-three residents there had 793 parcels among them, meaning that many had between 20 and 25 (with a mean of 12.6 per cultivator).[55] Even the smallest of parcels was cultivated intensively, by a single owner, taking advantage of the borders of irrigation canals or roads. For example, a Morisco owned one mulberry in a ditch between some houses in Adra. We can no doubt agree with Galán Sánchez's conclusion that this system was functional to the extent that some parcel was always in production.[56]

Muñoz Buendia and Diaz López have compiled interesting statistics on the Morisco minifundio in Pechina by aggregating irrigated and non-irrigated fields (measured in *tahullas* and *fanegas*, respectively) into four categories: microparcels (less than one *tahulla* or *fanega*), small parcels (1–5), middling (5–10), and large (10+) (Table 2). Here there was an absolute predominance of *minifundios* and even *microfundios*, with a pronounced pyramidal structure with a very few big owners at the top. Most Morisco proprietors (in Pechina at least) had estates of one parcel only.[57]

Although Barrios warns that we really don't know to what extent the agrarian landscape of the late sixteenth century was really Morisco or, in fact, a hybrid owing to borrowings from Christian practice, he and Burriel reach three general conclusions regarding

55 Angel Galán Sánchez, *Los mudéjares del Reino de Granada* (Granada, Universidad, 1991), p. 181.

56 Pedro Ponce Molina, 'Moriscos y repobladores. El paisaje agrario de Adra en la segunda mitad del siglo XVI', in *Almería entre dos culturas* (Almeria, 1990), II, 839–59, on p. 843; Galán Sánchez, *Mudéjares del Reino de Granada*, p. 183.

57 Antonio Muñoz Buendia and Julián Pablo Diaz López, 'Continuidad y cambio de la estructura agraria almeriense en la edad moderna: El caso de Pechina', in *Almería entre dos culturas. Siglos XIII al XVI*, 2 vols. (Almería, Instituto de Estudios Almerienses, 1990), II, 731–62, on pp. 750–1.

the nature of agrarian change and continuity across the process of *Repartimiento,* from a Morisco to a Christian landscape. First, the typical Morisco *minifundio* was attenuated as individual grants were generous. Second, there was initial equality among the settlers, according to social class, just as had been characteristic of the medieval *Repartimientos.* Third, the process of accumulation and fragmentation (together with the typical land market which, we have seen, arises after *Repartimiento*) favoured certain groups more than others and introduced inequalities. But, they go on to say, these changes were not accompanied by any significant change in the pattern of parcellisation. Since the rules laid down by the King specified that each lot had to have a mixture of parcels, the general dispersion and size of parcels was maintained.[58]

In Pechina, for example, the *minifundio* disappeared. Instead of the predominance of the *minifundio* and *microfundio,* 100 per cent of estates in Christian Pechina directly after the partition were large ones (10 *tahullas/fanegas* or more), although the estates were highly dispersed, rather than concentrated, following Crown policy.[59] For example, one Pedro Asnar, a settler from Valencia, received a house in Mojácar, three *fanegas* of *secano* in Campillo, ten more in Albolucar and more yet in different, named places; then four olive trees located on a *cañada,* six irrigated olives, eighteen irrigated mulberries on a terrace, plus twelve more in a different place; a *tahulla* and a half of irrigated land with olives and mulberries in Daimuca, and so forth. The antecessor of each discrete parcel is identified; thus in this case it would be possible to reconstruct rather exactly the pattern of dispersion of Morisco property.[60]

Cortijos

We noted in Chapter 2 the existence, alongside *alquerías,* of privately owned estates called *majshar/s.* In the *Repartimientos* of Seville, and many places in Granada these '*machars*' are usually rendered in Spanish as *cortijo.* González Jiménez presumes the early modern *cortijo* to have been a large estate formed in the Islamic period,[61] but the absolute coincidence between pre-conquest *majshar/s* and post-conquest *cortijos* is not easy to establish. Oliver Asín, noting the high

58 Barrios Aguilera, 'Repoblación', p. 627; Barrios Aguilera and Burriel Salcedo, *Repoblación de Granada,* pp. 51–2.

59 Muñoz Buendia and Diaz López, 'Continuidad y cambio', pp. 750, 752.

60 Juan A. Grima Cervantes, *La expulsión morisca y la repoblación cristiana de Turre (1570–1596)* (Almería, Diputación Provincial, 1988), pp. 171ff.

61 González Jiménez, 'Repartimientos andaluces', p. 111. In Seville, these estates frequently bore the names of important Arab families.

incidence of *majshar/s* in the Seville area, alongside a plethora of place-names ending in *-ena*, indicating a unit of property, presumes the Andalusian *cortijo* to have a Roman or pre-Roman ancestry.[62] Here again, we note the attraction of an older generation of Spanish scholars to the Sybiline lure of Roman 'origins'. The size and use of these estates were pretty much dictated by their physical characteristics and the possibility, or impossibility, of irrigating them. The word *majshar*, according to Oliver Asín (following Dozy), is most likely a Latinism, from *massaria*, and its use was virtually restricted to Andalucía and Mérida. In my view, the significance of this localised use has little to do with any putative continuity with Roman agriculture: in all regions of Al-Andalus, we have seen, the rural landscape comprised villages, generally arrayed in castral districts, and private estates which had different names in different regions and which were generally unirrigated.

Luna Díaz describes sixteenth-century *cortijos* in the Montes of Granada as the 'prototype of the dispersed habitat'. They all had in common their isolation and their lack of any separate administration or jurisdiction, although many such places were in fact hamlets of between three and fifteen houses. Most, however, had one to three houses. Those of one isolated house were typically occupied seasonally, while hamlet-style *cortijos* were permanently occupied. Moriscos owned smaller *cortijos* (less than 200 *fanegas*), whereas all of the larger estates belonged to Christians. Although only some had grist-mills, virtually all were dry-farmed for wheat.[63] Luna does not even mention possible continuity with the Muslim *machar*. The hamlet-style *cortijos* with multiple houses, which he calls *cortijadas*, were almost certainly the remains of depopulated Muslim *alquerías*, transformed into *cortijos*.[64]

A tree-growing society

The *Repartimientos*, particularly those of the fifteenth and sixteenth centuries, provide a very detailed picture of the distribution not

62 Jaime Oliver Asín, 'Maysar-cortijo: Orígenes y nomenclatura árabe del cortijo sevillano', *Al-Andalus*, 10 (1945), 109–26, on pp. 122–6; Vincent Lagardère, *Campagnes et paysans d'Al-Andalus* (VIIIe–XVe s.) (Paris, Maisonneuve et Larose, 1993), pp. 109–11.

63 Juan Andrés Luna Díaz, 'Repoblación y gran propiedad en la región de los Montes de Granada durante el siglo XVI. El cortijo', *Chronica Nova*, 17 (1989), 171–204, quotation on p. 176.

64 *Ibid.*, p. 189, and González Jiménez, 'Repartimientos andaluces' (1990), p. 107. On *cortijadas* in the Alpujarras, see Marie-Christine Delaigue, 'Mutations de l'espace villageois en Andalousie orientale. Effets immédiats et lointains de la Reconquête', *Mélanges de la Casa de Velázquez*, 26 (1990), 131–62, on pp. 143–4.

only of fruit and olive trees in Muslim-inhabited areas, but also mulberries, the raising of which was lucrative because of their role in the silk industry, and then native trees which yielded nuts, wood, or bark. This was an economy in which tree crops played a larger role than they had before or after the last periods of Islamic settlement, probably because in an agrarian world of *minifundios* they offered the promise of a cash-crop that was economical of space. In thirteenth-century Christian Mallorca, arboriculture accounted for 30.35 per cent of agricultural land use, irrigated field crops, 32.13 per cent, and vineyard 30.37 per cent (it was impossible to figure cereal production). Inasmuch as grapes were introduced by the Christian, Muslim agriculture there was divided between arboriculture and irrigated huertas.[65]

Furthermore, the Muslims practised a highly intensive form of *coltura promiscua* in which trees which elsewhere were not irrigated were intercalated with irrigated field crops to increase their yields. The *Repartimiento* of Almería records no less than 18,659 trees including 4,714 pomegranates, 4,284 figs, 2,649 mulberries, 1,424 cherries, and 1,422 apples.[66] Fruit trees were grown on the borders of fields, both dry-farmed and irrigated, although on the latter more trees and more species and varieties of trees were found.[67] As specified for the village of Pinillos in the Vega of Granada in 1500:

> Here are the said 5750 *marjales*[68] of field and vineyards of unirrigated and irrigated land, in which there are olives and mulberries, not including many other trees which were not counted, such as almonds, cherries, carobs, mulberries, figs, and other fruit trees which are in the said vineyards and dry and irrigated fields. Because (the trees) are not surveyed, but rather all mixed up, one here and another there, they cannot be called orchards and one has to grant them to whomever merits a lot of trees on the borders of dry or irrigated fields, as the lot falls to him.[69]

In the vega of Granada, fruit trees were generally irrigated, olives not.[70] In Almería, on the other hand, olives were frequently irrigated. A number of entries in the sixteenth-century *Repartimiento* of

65 Soto i Company, 'Repartiment i "Repartiments"', p. 38.
66 Cristina Segura Graiño, *El Libro de Repartimiento de Almería* (Madrid, Universidad Complutense, 1982), pp. 66–7.
67 Antonio Malpica Cuello, 'Repoblaciones y nueva organización del espacio en zonas costeras granadinas', in Miguel Angel Ladero Quesada, ed., *La incorporación de Granada a la Corona de Castilla* (Granada, Diputación Provincial, 1993), pp. 513–58, on p. 546.
68 Measure of area, 1 marjal = 528 sq m.
69 Pedro Hernández Benito, *La vega de Granada a fines de la Edad Media según las rentas de los habices* (Granada, Diputación Provincial, 1990), p. 283.
70 *Ibid.*, pp. 74, 77.

Turre, Almería, water rights were assigned to specific olive groves,[71] and when lots were enumerated, the grant specified whether olives (and mulberries) were irrigated or not.[72] Likewise in Yunquera (Málaga), settlers of formerly Morisco villages received 5.5 *aranzadas* of 'good' grape-vines (plus 2.5 of 'old'), two *fanegas* of irrigated land.[73]

On the southern coast of Almería, in the vicinity of Salobreña, there was a characteristic mixture of the exotic – like bananas – and the familiar, like the oak, explicable, according to Malpica, 'by the juxtaposition of intensive irrigation and the utilisation of the mountains'.[74] In the Andarax Valley, Old Christians living there in the 1570s were asked what 'poplars, canebrakes, pines, chestnuts, and oaks there were in that district (*taha*)'; they reported that the poplars had been cut down but that many oaks remained in common pasture land.[75] Figs, which were extremely numerous in Andalucía, were typically associated with grape-vines.[76]

There were two varieties of mulberries: the *moral* (*Morus nigra*) and the *morera* (*Morus alba*): the latter was an Italian import, easier to grow, but not as high yielding. The surveyors measured mulberry groves by the square foot, the same measure used for olive trees, and the productive capacity of the trees was recorded in 'ounces of silk production' (*onzas de criar seda*).[77] Because the silk industry underwent a boom in the sixteenth century, the distribution of trees changed over time. Thus in Pechina 71.6 per cent of trees were olives in 1498, 21.8 per cent figs, and 6.6 per cent mulberries (*morales*). In 1573, that distribution had changed to 58 per cent olives, 32 per cent mulberries (mainly *moreras* now), and 9.9 per

71 Grima Cervantes, *La expulsión morisca*, p. 113: 'un olivar . . . con dos días de agua' and another with half a day of water; p. 114: 'un olivar . . . con un día de agua este bancal y el de abajo'; also eleven olive trees with specific water assigned from a tank.

72 *Ibid.*, pp. 171–2: grant to Pedro Asnar of Valencia includes a house in Mojácar, four olive trees on a *cañada* (unirrigated by implication), six irrigated olive trees, eighteen irrigated mulberries, etc.

73 Cabrillana Cizar, *Yunquera*, p. 122.

74 Antonio Malpica Cuello, 'La implantación castellana en la tierra de Salobreña. La alquería de Benardila', *Revista del Centro Histórico de Granada y su Reino*, 2nd epoch, 3 (1989), 199–213, on p. 204. Note the familiar constellation: an irrigated *alquería* with a Beni- name, in the orbit of a *ḥiṣn* (Salobreña). On the cultivation of bananas, see Andrew M. Watson, *Agricultural Innovation in the Early Islamic World* (Cambridge, Cambridge University Press, 1983), pp. 51–4.

75 Pedro Ponce Molina, *Repartimiento de Dalias/El Ejido* (Almería, Quash/Tierras de Almería, 1984), p. 35.

76 Malpica Cuello, 'Repoblaciones y nueva organización', p. 546.

77 Manuel Espinar Moreno, 'Los árboles en las tierras de Cantoria. Suertes confeccionadas y reparto', *Roel*, 6 (1985), 139–69; Hernández Benito, *La vega de Granada*, pp. 43, 75.

cent figs.[78] Patterns of change differed from place to place, however. In the Sierra de Filabres, the Muslim cultivation system was kept intact, except that here the mulberry yielded to the olive.[79]

Houses too small: *casas moriscas*

The *Repartimientos* have left a record of the appropriation by Christians of Muslim houses, generically referred to by the *repartidores* themselves as *casas moriscas*: fifteenth-century descriptions typically use the diminutive: *casas pequennas, una casa pequenna, casiellas*[80] – and are not limited to houses, but rather a characteristic of all vernacular Muslim buildings: a house that had been an *alfondiguilla* – a little warehouse, another which had been a little mosque – *mezquitiella*. When Christians entered a Muslim town with intent to take up residence there, the scale of the cityscape struck them as too small. As a result two or more houses were put together. Frequently settlers were granted a pair of houses, with the anticipation that out of two *casas moriscas* a proper Christian dwelling could be fashioned. The typical Muslim house had a living room/bedroom on the ground floor – called the *palacio*, a kitchen, and corral. Then on the second floor was a storage place for grain, hay or other agricultural produce, called the *algorfa*.[81]

In an archaeological reconnaissance of houses in Mudéjar villages in the province of Alicante deserted prior to the sixteenth century, Torró and Yvars found small houses with rooms on the first floor measuring from 9 to 12 m long by $2\frac{1}{2}$ m wide. There seems to have been only one room, inasmuch as no sign of partitioning was uncovered. Most of these houses had an *algorfa* above and a roof sloping down towards the corral. These houses were not built with *tapia* (rammed earth) – the material of choice for many *ḥuṣūn* and other fortifications. The same technique was used – a method of building in regular courses using a wooden mould

78 Muñoz Buendia and Díaz López, 'Continuidad y cambio', p. 761.

79 Juan García Latorre, 'La pervivencia de los espacios agrarios y los sistemas hidraúlicos de tradición andalusí tras la expulsión de los moriscos', *Revista del Centro de Estudios Historicos de Granada y su Reino*, 2nd epoch, 6 (1992), 297–317, on p. 309.

80 Manuel González Jiménez and Antonio González Gómez, *El Libro del Repartimiento de Jerez de la Frontera* (Cádiz, Instituto de Estudios Gaditanos, 1980), pp. 10–12, 24, 103.

81 López de Coca Castañer, *La tierra de Málaga*, p. 28. Cf. examples from the *Repartimiento de Bezmiliana* (1492), *ibid.*, pp. 473–6: a house with a good, roofed *palacio* and another with an *algorfa*, fallen in and with a corral next to it; a house with a *palacio*, kitchen, and storeroom, covered with a terrace; a house with a *palacio* and two kitchens; another with two small *palacios*, etc.

(*tapiera*) in which the building material was packed down – but instead of earth house-builders used either mortar with a fine aggregate of pebbles (*tapial de mortero*) or a mixture of large rocks and mortar (*tapial de mamposteria*).[82]

In towns, small houses were less predominant. For Jerez, González Jiménez found that *casas moriscas* accounted for one-quarter of the 2,585 houses recorded (627), while 64 per cent (1,634) were medium-size, and only 3 per cent (87) large.[83] Presumably, the poorer and smaller the settlement, the greater the predominance of the *casa morisca.*

Casas moriscas were so small that grantees frequently changed their use: thus a house in Ronda was converted into a stable;[84] a winery (*bodega*) in Jerez was described as consisting of two little pairs of *casas moriscas* with some stores; a large corral which had been three pairs of *casas moriscas*; another which had ten pairs of Muslim houses; a *bodega* which had been a mosque, and so forth.[85] From the late fifteenth century on, due to the continual abandonment of houses, many were described as being in poor condition: *malas casas* in El Burgo (in 1492) or in Bedar (1575), where 50 per cent of all houses were fallen in (*caídas*), without roofs (*destechadas*), or torn down (*derribadas*).[86]

In their archaeological reconnaissance, Torró and Yvars found evidence that, over time, a second little room would be added on the ground floor, either forming an L with the original room or else built out on the corral side. The division of the house into two

82 Josep Torró and Josep Yvars, 'La vivienda rural mudéjar y morisca en el sur del País Valenciano', in *La casa hispano-musulmana: Aportaciones de la arqueología* (Granada, Patronato de la Alhambra, 1990), pp. 73–97. On *tapia* as building material, see Thomas F. Glick, 'Cob Walls Revisited: The Diffusion of Tabby Construction in the Western Mediterranean World', in B. Hall and D. West, eds., *On Pre-Modern Technology and Science: Studies in Honor of Lynn White, Jr* (Los Angeles, 1976), pp. 147–59; Pedro López Elum, *La alquería islámica en Valencia: Estudio arqueológico de Bofilla, siglos XI a XVI* (Valencia, privately printed, 1994), pp. 85–92; and André Bazzana, *Maisons d'Al-Andalus: Habitat médiéval et structures du peuplement dans l'Espagne orientale*, 2 vols. (Madrid, Casa de Velázquez, 1992), I, 76–82.

83 González Jiménez, 'Repartimientos andaluces' (1990), p. 111.

84 Acién Almansa, *Ronda y su serranía*, II, 182.

85 González Jiménez, *Repartimiento de Jerez*, pp. 16, 41, 79, 83. On conversion of mosques to houses, houses to stables and *bodegas* generally, see González Jiménez, 'Repartimientos andaluces' (1990), p. 116.

86 Rafael Benítez Sánchez-Blanco, 'El Repartimiento de El Burgo (Málaga), 1492: Estudio de su estructura agraria', in *Homenaje al Dr D. Juan Reglá Campistol* (Valencia, Universidad, 1975), I, 227; and Juan Jesús Bravo Caro, 'Vivienda y tierra de riego en Bedar en el momento de la expulsión de los moriscos', in *Almería entre dos culturas*, 2 vols. (Almería, Instituto de Estudios Almerienses, 1990), II, 863–76, on p. 870.

equal parts suggests that the segregation of a conjugal or strictly nuclear family out of the original domestic group did not require dividing the parcel in two, but rather the building of a new cell for a second nuclear family, at the expense of the common corral or patio. Inasmuch as these houses postdate the Christian conquest of the 1240s the authors ask whether feudal control hastened the end of the joint family or clan organisation of rural Muslim life.[87]

Remodelling the *Ḥuṣūn*

The implantation of feudal norms necessarily entailed the adaptation of the system of *ḥiṣn/qarya* complexes to them. I have previously indicated that this process sheds some light on how that system functioned, in the absence of Arabic documentation: what the Christians added suggests what was *not* present in the pristine complex.

The *ḥuṣūn* of the Islamic period underwent a complete transformation with the Christian conquest, not only in the organisation of each castle but also in the structure and function of the network of castles. The castle/refuge was transformed into a seigniorial castle and endowed with a new function: the domination of the castral zone, with its *alquerías,* politically, socially and economically. In Valencia, no less than 150 *ḥuṣūn,* now called *castra* together with their Muslim communities (*aljamas*) still led by their shaykhs, called *vells,* were granted out to Aragonese lords.[88] Moreover, archaeological evidence documents the functional change in *castra: celoquias,* which had been rather simple residences (indeed sometimes only temporary ones) for *qā'id*/s, were now upgraded and remade into permanent residences that a noble or his castellan could inhabit with his household, including whatever retainers were required for administering the castral district. In the case of the most elemental refuge, such as the *ḥiṣn* of Perputxent, which had an *albacar,* but in so far as we know, no *celoquia,* the new Christian lord superimposed a *castrum* on top of it. This particular *ḥiṣn* had had no strategic or administrative function; under the Christians it was transformed into a feudal castle. Elsewhere, as in the castles of Castells and Penáguila, sub-castral settlements of Christian settlers were formed outside the castle walls, with mixed success. Better luck was had by

87 Torró and Yvars, 'Vivienda rural mudéjar', p. 79.

88 Pierre Guichard, 'Oriente y occidente: Población y sociedad', in his *Estudios sobre historia medieval* (Valencia, Diputación Provincial, 1987), pp. 105–32; Guichard, 'El impacto de la Reconquista en la sociedad musulmana', in *Historia del Pueblo Valenciano* (Valencia, Levante, 1988), p. 232.

new settlements built in the *albacar/s* of former *ḥuṣūn*, such as one *in albacharo de Biar* (1287), another at Castalla, where 100 newcomers found housing *in circuito castro*, and another at Cárcer (1281). This was a defensive settlement strategy in an area of great danger for the small Christian settlements and which saved the cost of building walls around new settlements. In Guadalest, repairs to the old *ḥiṣn* led to a new settlement within the *albacar* and the conversion of the old *celoquia* into a proper feudal castle.[89]

Settlement patterns in the immediate environs of *ḥuṣūn* were also altered. In Cocentaina and Xàtiva Christians did not occupy the old Muslim urban space high up near the *castrum*, but established themselves lower down in the huertas. In Denia there was an inverse pattern, where Christians settled in the *albacar* of the *castrum*, leaving the Muslim settlements below intact. Christians, according to Torró, typically liked to establish new towns rather than re-use Muslim urban structures which were not to their liking. In Castelló, the Muslim *ḥiṣn* of Castilione had controlled a district of *alquerías* in the irrigated huerta, including Benimahomet and Benirabe. Settlers were granted lots in the latter which subsequently disappeared in the street plan of a substantially new town, Castelló, which had nothing to do with the old *ḥiṣn* of the same name.[90]

In Andalucía, virtually the same phenomenon has been described: the *ḥiṣn* of Gibralfaro, for example, was made smaller by the Christians, who placed a small *alcázar* in one corner of the site, in conformity for the feudal requirement that an *alcaide* and small garrison reside there.[91]

Irrigation: learning the Muslim system

The third element was irrigation. Here there is a paradox: even in those places where there was retrenchment of irrigated garden agriculture and where its share of the total sector lost out to cereals and vineyards, the distribution regimes were kept intact. In some

89 André Bazzana, Pierre Guichard and José María Segura Martí, 'Du *ḥiṣn* musulman au *castrum* chrétien: Le château de Perpunchent (Lorcha, province d'Alicante)', *Mélanges de la Casa de Velázquez*, 18 (1982), 449–65. On varieties of *ḥiṣn* transformations, see Josep Torró, 'El problema del hábitat fortificado en el sur del reino de Valencia después de la segunda revuelta mudéjar (1276–1304)', *Anales de la Universidad de Alicante. Historia Medieval*, 7 (1988–89), 53–81. Pedro López Elum doubts that the Christian castle of Perputxent had been a *ḥiṣn* in Islamic times: 'Castelología y cerámica medieval: Propuestas arqueológica y consideraciones metodológicas', *II CAME*, I, 231–43, on pp. 232–8.

90 Torró, *Poblament i espai rural*, pp. 73–5.

91 Miguel Acién Almansa, 'Recientes estudios sobre arqueología andalusí en el sur de Al-Andalus', *Aragón en la Edad Media*, 9 (1991), 355–69, on p. 366.

cases, instructions regarding irrigation arrangements were included in the *Repartimiento* itself, as in this passage from that of Orihuela in 1272:

[It is ordered] that all the property owners of Orihuela, not only those who have grants, but also the other residents, be made to clean and repair the drainage ditches and the large and small irrigation canals of the territory of Orihuela, so that the water might flow without impediment just as it flowed in the time of the Moors. And let them apportion the water by *tahullas* to each one as he had it, just as they lawfully had in the time of the Moors. We order them to seize the properties of those who do not wish to obey and give them to whoever will uphold custom and neighbourly duty. And if anyone should force the irrigation officers [*acequieros*] to give them water, let them forfeit their persons and everything they have to the king.

Furthermore we order that no one dare to plant grapevines in those irrigated places that are for cereals. Those who so plant, let it be taken for the king. [92]

It is particularly clear in this passage that it was royal policy 'to uphold custom'; in this case, that of the antecessor Muslim irrigators.

In other instances, special inquests were held, such as the one conducted by Peregrín de Atrosillo in 1244, after the conquest of Gandía, in which Atrosillo quizzed Muslim irrigators regarding the apportionment of water among the various communities irrigating from the Serpis river.[93]

Operating procedures thus displayed considerable continuity, even though the tribal model of governance that everywhere prevailed in the irrigation communities of the rural Islamic world had to be replaced. The Christians simply appropriated a proximate model of collective work organisation, that of the craft guild, and applied it to the administration and governance of autonomous irrigation communities. As a result, irrigation communities, whether Christian or Mudéjar, came to resemble guilds of irrigators. The irrigators collectively formed the commons of the canal. They met once yearly to discuss mutual affairs and elect officers who were endowed with considerable discretional authority to act on routine matters. The general officers of the commons, syndics, and *veedores*, were the same as found in any guild. Communities thus constituted 'juridical personalities' and were able to make use of the royal court system, both for the solution of internal problems and of conflict with neighbouring irrigation communities. Medieval Spaniards'

92 Torres Fontes, ed., *Repartimiento de Orihuela*, p. 51.

93 Roque Chabás, *Distribución de las aguas en 1244 y donaciones del término de Gandía por D. Jaime I* (Valencia, Francisco Vives Mora, 1898).

love of litigation, when compounded with irrigation's potential for generating conflict and Roman law's standards of evidence, produced thousands of pages of testimony by expert witnesses, many of them Muslims. It is this kind of documentation that makes it possible to reconstruct pre-conquest Islamic irrigation systems and track their change over the cultural transition.

Strict continuity with Muslim practice was not displayed everywhere, however. In Bizar, a Muslim *alquería* near Guadix known by the Christians as Policar, a dual system of water rights arose in the 1490s, a Muslim one dating to a twelfth-century apportionment of water, and a Christian one based on a royal decree of 1494 establishing guidelines for water officials (*alcaldes de aguas*). Litigation ensued between the two groups of irrigators.[94] But this kind of dispute was quite rare in post-conquest irrigation.

In Motril, Muslim communal control of irrigation was abolished and usurped by the Christian town council which, in 1510, appointed a *mayordomo* to oversee irrigating on the town canal, replacing the *viejos honrados* (honoured elders, that is, *shaykh*/s) of the Islamic system.[95] Something similar happened in the city of Almería where the Church had assumed the ownership of the religious trusts (*habices*) and therefore was obliged to maintain the irrigation canals of the city, according to a decree of 1492. The Church, however, was unable to keep the town canal in repair (to the prejudice of the townspeople, whose residential cisterns were supplied from the town canal), and in 1503 ownership of the water was transferred from the Church to the town council. The struggle between the Church and the town over water lasted until the eighteenth century.[96]

The instructions to officials executing both *apeos* and *repartimientos* after the expulsion of the Moriscos from Granada were frequently explicit regarding irrigation. Thus the A*peo de Alfacar* (1571) was to include 'the water which they have, how they irrigate with it, from what rivers, by which canals . . . from which river is it derived, and in what order each is accustomed to irrigate, and what the water rights of the Moriscos were, by day and by night'.[97] This kind

94 Manuel Espinar Moreno, 'Bizar: una alquería musulmana y el paso al dominio cristiano (siglos XII–XVI)', in *Andalucía entre Oriente y Occidente (1236–1492)* (Córdoba, Diputación Provincial, 1988), pp. 707–18.

95 Manuel Domínguez García, 'La acequia de riegos de Motril y las ordenanzas de 1561', *AZA*, II, 951–68, on p. 958.

96 Jesús M. López Andrés, 'La intervención de la iglesia de Almería en la administración de las aguas del abasto del común de la ciudad', *AZA*, II, 863–73; Francisco Andújar Castillo, 'Adaptación y dominio del agua. La vega de Almería en el primer tercio del siglo XVII', *AZA*, II, 1087–99, on p. 1096.

97 Barrios Aguilera and Burriel Salcedo, *Repoblación de Granada*, pp. 198, 200.

of *apeo* was something like what in American water law is called an adjudication procedure. In the course of the *apeo*, the official must oblige each irrigator to prove or establish what his water right is. In Loaysa's *apeo* of the Aindamar irrigation canal of Granada in 1575, he was charged with determining 'the water which belongs to each of the houses and estates in property [= water right], and the order and form which those who were owners of the said estates followed in enjoying it'. An abuse had arisen when the lessees of the canal had taken advantage of the ignorance of the new Christian proprietors; thus Loaysa's orders stipulated that there be a return to 'the order and custom prevailing in the time of the Moriscos', in this case a fixed rotation from April to October, with different classes of right-holders, according to the amount of water each was permitted.[98]

Likewise a Royal Provision of 1571 covering *repartimientos* generally specified that broken or disturbed canals be repaired so that settlers could irrigate without confusion or conflict. The *Repartimiento of Irtrabo* (1571) specified that settlers must clean canals, direct water for irrigation and 'observe the ordinances of the chief village in the district in regard to the way of irrigating'.[99] Some instructions for Partition of Water of the Plain of Andarax (1574) specified that an 'intelligent person' be named to appraise the irrigated fields of each place and declare the days when each district irrigates, 'taking into consideration the amount of land that each canal irrigates . . . so that all might irrigate equally'.[100] That is, the document recognises that proportional distribution was the rationale of the system of allocation of water and that a high value was placed on equality. Some *Repartimientos* gave unusually specific details regarding the mode of water distribution: in that of Casarbonela, owners of irrigated parcels were obliged to shut off access of their canals to the main channel once they had irrigated.[101] Such a provision (a standard practice) would seem directed towards settlers previously unfamiliar with community irrigation.

Information from the Lower Andarax in the sixteenth century illustrates what happens to the successor of *ḥiṣn/qarya* irrigation, in this case in the *taha* of Marchena. In 1560, in defiance of ancient custom, the city of Almería seized water that had always belonged to upstream *alquerías*, causing indignation among the Moriscos

98 Barrios Aguilera, *Moriscos y repoblación*, pp. 138–40. Cisterns for drinking water had the highest priority.
99 *Ibid.*, pp. 219, 227.
100 Ponce Molina, *Repartimiento de Dalias*, p. 141.
101 López de Coca Castañer, *Tierra de Málaga*, p. 38.

residing there. (The city naturally came into conflict with upstream irrigators because its drinking water came from fountains supplied by *cimbras* from the Andarax river.) In irrigation turns (*tandas*) implemented in 1502 and 1572, it was stated that when water is short the right belongs to the place immediately upstream of where the water is lacking. Here we see even at this rather late date the working out in practice of a principle of Berber customary water law, favouring upstream users, the rationale for which is obvious in times of drought. This system remained intact as long as there were Moriscos there to defend their ancient rights. When they left, the wealthy – citizens of Almería who owned haciendas in this area – were able to manipulate the system to deny water to new settlers.[102] In this period, Almería made a transition from a Muslim commercial city to a kind of Christian 'Agro-town', in which resided new settlers with rural land grants. Land was taken out of irrigation and converted to cereals, in keeping with the prior agricultural experience of new settlers.[103] Finally, in 1584, the King complained that many properties had been ruined because the administration of water had decayed and irrigation canals were no longer built and maintained as had been the custom in 'the time of the Moriscos'. The immigrants, totally unfamiliar with Andalusi irrigation traditions, even replaced the Arabised terminology of microsystem accoutrements with Latinisms: the *alberca* becoming a *balsa* [tank], the *aljibe* [cistern], an *estanque.*[104]

It has become a leitmotif in the historiography of Morisco Granada that the Moriscos retained communal ownership of water.[105] But this is certainly not true in most places in Granada, if compared to the description of communal ownership by a *qawm* in eleventh-century Tunisia, as described in Chapter 4.[106] In sixteenth-

102 Antonio Gil Albarracín, 'Los regadíos del bajo Andarax durante el siglo XVI', *AZA*, II, 969–80; Andújar Castillo, 'Adaptación y dominio del agua', pp. 1090–2.

103 Land transaction grants change in the course of the sixteenth century from *regadío y un poco de secano* to *tierras de panllevar.*

104 María Isabel Jiménez Jurado, 'La ruralización de Almería en el siglo XVI. Problemas económicos derivados de la irrigación de la tierra', *AZA*, II, 1005–15.

105 See typical assertions by Malpica Cuello, *Turillas*, p. 26; and Juan Jesús Bravo Caro, 'Vivienda y tierra de riego en Bedar en el momento de la expulsión de los Moriscos', in *Almería entre dos culturas* (Almería, Instituto de Estudios Almerienses, 1990), II, 863–76, on p. 872: 'todos goçavan della (water) respecto la tierra que cada uno tiene', which is a statement of the principle of proportionality, not of ownership.

106 In the Sierra de Filabres, water was said to have been 'common among the residents who have no property in it nor pay any tribute for it' ('común de los vecinos dellas que en esto no abía propiedad ni pagaban por ello tributo ninguno') (García Latorre, 'Pervivencia de los espacios agrarios', p. 299). In the *Repartimiento* of Montejicar, specified units of water were common to all ('agua arriba va

century Granada, the organisation of irrigation remained communal, by its very nature, and some measure of tribal-style governance was maintained, but the rights structure had been steadily privatised from some indeterminate time in the Middle Ages. The *Libros de Habices*, in particular, show that individual properties were alienated with specific *dulas* – in this case, hours – of water. Whether such ownership was usufructory or proprietary is irrelevant to the point.[107]

There is an interesting debate over the survival of Muslim irrigation practices across first the fifteenth-century conquest and, later, the expulsion of the Moriscos, and whether the new Christian settlers displayed disinterest, ignorance, or lack of ability in irrigation agriculture. Thus Malpica Cuello asserts that in post-conquest Granada of the late fifteenth century, irrigation was not adapted to the Castilian economy: 'Communal uses of water [and] preservation of irrigation systems are almost impossible for a Castilian population to maintain.'[108] Similar is Galán Sánchez's assertion that Christian settlers in the fifteenth-century vega of Granada broke the delicate juridical and social equilibrium required to keep the multiplicity of small canals functioning.[109] These kinds of blatant *obiter dicta*, with vague appeals to a supposed anti-irrigation *mentalité* among Castilians, have been proven wrong. What has been demonstrated, by Galán Sánchez and others, is that Christians did not always understand how irrigation was to be administered and in some cases new institutions had to be devised. Above I have discussed the replacement of tribal governance of autonomous irrigation communities by a craft guild model. In Granada, the municipal administrative system had to be altered. The canals had been overseen in Nasrid times by *amīn*/s (Castilian, *veedores*) remunerated with shops and rents belonging to the *habices*. Conflict over water

comun a todos y desde alli la mesma cañada arriba derecho de agua comun a todos'); María José Osorio Perez and Rafael G. Peinado Santaella, 'El Libro de Repartimiento de Montejicar (1527): Comentario y edición', *Revista del Centro de Estudios Históricos de Granada y su Reino*, 2nd epoch, 4 (1990), 71–112, on p. 104.

107 See a list of privately-owned hours of water in an *alquería* in Manuel Espinar Moreno, 'Estudio sobre propiedad particular de las aguas de la Acequia de Jarales (1267–1528). Problemas de abastecimiento urbano y regadíos de tierras entre las alquerías de Abrucena y Abla', *AZA*, I, 247–66, on p. 261.

108 Malpica Cuello, 'Poblamiento del Reino de Granada', p. 390.

109 Galán Sánchez, *Mudéjares del reino de Granada*, p. 172. The King had ordered Muslim water customs to be written down in Castilian. Those charged with the task simply settled matters in their own way, using precedents from Valencia and Murcia, while ignoring 'agricultural traditions' of Granada (*ibid.*). Of course, there is nothing necessarily dysfunctional about tinkering with institutions.

was endemic in the post-conquest city of Granada of the 1490s because canal water was used not only for irrigation but for supplying household cisterns. Therefore there were an unusually large number of right-holders. The Crown accordingly created a water tribunal to compose conflicts and, soon after, a Veedor de las Aguas to inventory the system and its distribution customs. Galán Sánchez sees this adjustment as contributing to the degradation of the system, but there is no inherent reason why new rulers should not develop functional administrative instruments congenial to their own institutions and jurisprudential traditions.[110]

In a study of the eastern half of the kingdom of Granada (comprising the bishopric of Almería, which was 90 per cent Morisco in 1568), García Latorre lists four generalised aspects characterising rural communities of Moriscos: first, the predominance of irrigation; second, the importance of arboriculture, especially the cultivation of mulberries; third, the absolute predominance of small property; and fourth, 'the existence of communitarian practices and social relations respecting the use and maintenance of irrigation (water as collective property tied to land, collective responsibility in maintenance of hydraulic systems)'.[111] This basic scheme, he asserts, was not seriously modified by the intrusion of the Church (as successor holder of *habices* as well as of its own substantial estates), nor the creation of many, generally modestly sized, seigniorial estates. Even in the city of Almería, where as we have seen, Morisco irrigators had been dispossessed, the old system of irrigation was kept intact as *habices* passed into the hands of the Church and the municipal irrigation system to the city council. In most areas, Moriscos continued to control mills, even as the new Christian lords obtained monopolies over ovens. This system of what García Latorre characterises as 'low feudal pressure' was consciously implanted by the Castilian authorities and had two significant outcomes: the first was that Morisco peasant communities retained a 'high degree of control' over local economic processes and labour and second, the State did not permit the establishment of true rent fiefs to compete with the peasantry in the functioning of local economies. The Crown sought to preserve for itself the right to exploit the Morisco population.[112] In so far as irrigation is concerned, of course, the bolstering of local control by Crown policy was a prime mechanism in preserving those systems intact and functioning.

Some authors have argued that Castilian settlers after 1568

110 Galán Sánchez, *Mudéjares del reino de Granada*, pp. 175, 177–9.
111 García Latorre, 'Pervivencia de los espacios agrarios', p. 298.
112 *Ibid.*, pp. 301–2.

were dry farmers and were unable to adapt to Morisco irrigation techniques. Thus, Bernard Vincent based such a conclusion on an analysis of forty-two pueblos in Almería, in which the canals were well maintained in twenty-seven; still he concluded that the traditional system was dismantled.[113] But Latorre argues there is no empirical evidence for such a conclusion. He studied ten villages of the Sierra de Filabres from Vincent's own sample, taking a closer look. In seven, the canals were repaired and operating. Only in one instance had the irrigation system deteriorated, but for social, not technical reasons.[114] What actually took place after the expulsion was a massive boom in irrigation, particularly of mulberry trees. Even though many of the new settlers were from Murcia and Valencia and might have already been irrigators, Latorre doubts that the origin of settlers explains the success or failure of post-expulsion irrigation. The key variable was whether the social norms that encouraged and regulated Morisco access to water (and supported local peasant control) had survived or whether some new set of norms was implanted.[115] In order to guarantee the continued success of the silk industry, the Crown ordered that agricultural practice remain the same as it had been in the time of the Moriscos. The social structure created by the resettlement was, just as its late medieval precursor whose norms it followed, egalitarian, which favoured a 'soft' takeover of hydraulic structures, rather than an abrasive one. The inspection of 1593 was explicitly designed to stop any feudal intromissions that might impede the maintenance and functioning of irrigation systems; *alcaldes* were ordered to clean and repair irrigation canals and charge irrigators on a pro rata basis. Thus the town council of Abrucena guaranteed collective responsibility for the maintenance of the canals, as representative of the irrigators: 'because this is the way of ancient custom . . . whereby this place has its irrigation water apportioned . . . by equal shares, giving to each water right (*suerte de aguas*), that which is apportioned by the clock'.[116]

Were Andalusi irrigation systems feudalised?

In the twelfth and thirteenth centuries, Catalan conquerors made a number of attempts to feudalise Andalusi irrigation systems, or at

113 Bernard Vincent, 'La société chrétienne almeriense et les systemes hydrauliques. Quelques propositions de travail', *AZA*, I, xciii–cix.
114 García Latorre, 'Pervivencia de espacios agrarios', p. 305.
115 *Ibid.*, pp. 307–8.
116 *Ibid.*, pp. 312–14.

least to make feudal grants inappropriate to the customary operating procedures of the canals in question. In all documented cases, such experiments failed and the canals were either municipalised, returned to communal control or if not, continued to function ineffectively. In Lleida, Pere Ramon Çavacequia was declared *sequier* (canal officer) of the Segrià canal for life, and apparently owned the canal. In 1190, apparently after conflict with the irrigators over the *sequiatge* – a user fee paid for upkeep of the canal which Pere Ramon had turned into a feudal tax – the financial structure of the canal was normalised, a prelude to the city's purchase of the canal in 1213.[117] The *sequiatge* was the most obvious pressure-point between the community of irrigators and feudal lords, particularly if the latter mistook this user fee for another seigniorial tax, or tried to convert it into one. In Muslim times, this fee was almost certainly paid in money, but Pere Ramon collected his in kind (grapes and grain). Similarly, when James I built a new canal – the Royal Canal of Alzira – he too appointed a *sequier* for life and later encumbered the *sequiatge* of Alzira with a money fief of 2,000 *solidi* annually to a French magnate. This payment, together with the *sequier*'s 500 *solidi* salary, indicates that the *sequiatge* was perceived in the thirteenth century as something more than a mere user fee. Nevertheless, this official too, like his counterpart in Lleida, ended up as a municipal employee (although not until 1393).[118] In the Huerta of Valencia, King James had retained one of the eight canals of the city – that of Moncada – as his personal possession; the rest he granted outright to the city. Moncada he granted to a magnate, Nicholas of Vallvert, but upon the death of the latter some thirty years later the King returned the system to its users, 'upon payment of a stiff privilege fee of 5,000 solidi'.[119] James's handling of these two 'royal' canals appears to have constituted an exception to the general rule, which he himself had enshrined in the *Furs* of Valencia, that water was public and its allocation for irrigation should proceed according the custom prevailing 'in the time of the Saracens'.

In Mallorca the King had caused havoc with the Síquia de la Vila, the main irrigation canal of the city of Palma. The King himself controlled water rights which he granted out to magnates,

117 Francisco Javier Teira Vilar, *El régimen jurídico de aguas en el llano de Lérida (siglos XII a XVIII)* (Barcelona, Universidad, 1977), pp. 97–101. *Çavacequia* is derived from Arabic, *Ṣāḥib al-sāqiya*, 'master of the canal'.

118 Robert I. Burns, *Medieval Colonialism: Postcrusade Exploitation of Islamic Valencia* (Princeton, Princeton University Press, 1975), p. 124; Tomás Peris Albentosa, *Regadío, producción y poder en la Ribera del Xúquer (La Acequia Real de Alzira, 1258–1847)* (Valencia, Generalitat Valenciana, 1992), pp. 31–7.

119 Burns, *Medieval Colonialism*, p. 126.

apparently violating the principle of proportional division of water underlying the original design of the system, whose functioning was further compromised by the royal authorisation of the construction of numerous new mills on the same canal. In 1239 water irrigating fields beyond the walls of the city was divided into two blocks of time: half was reserved for public use, the other half for the magnates and the Church.[120] This arrangement, not unsurprisingly, set in motion a chronic state of conflict over water rights. The monastery of La Real, granted far more water than its proportional share directly after the conquest, finally yielded to pressure (in 1310) and exchanged its right to one-third of all the water on one day weekly in return for an annual payment of 125 *morabetinos* and the right to keep a small delivery pipe open permanently. Water was sold illegally and the canal became less and less able to meet the needs of irrigators during periods of drought. Finally, the royal governor, pressured by the jurates of the city, initiated an adjudication procedure in 1381, forcing water users to produce their titles. Between 1230 and 1381, no less than fifty-five royal grants of water had been made. The governor responded by closing and reducing turn-outs, shutting down illegal diversions, and cancelling illegitimate turns – an act which stimulated an avalanche of suits.[121] Here is an example of the Crown's attempt to capture the high returns from irrigation agriculture directly by imposing a *cens* in return for the grant of water.

The cases of Lleida and Palma were not necessarily representative of royal policy, which appears to have been capricious, to say the least. After the conquest of Tortosa, villages were granted a measure of collective juridical authority: 'This village justice seems primarily to have concerned matters relating to irrigation or apportionment of running water; at any rate, in the region of Tortosa in 1165, the first public communal agent whose title is known to us was the *çavacequia*, the water-bailiff of the community of Horta.'[122]

There were certainly instances of feudal lords expanding irrigation in areas where Muslims had previously irrigated. The Cistercian monastery of Valldigna built a number of new secondary canals from the Alfandec river in the second half of the thirteenth century, and a grant of 1290 refers to a parcel in the lower part of

120 Lluis Tudela Villalonga, *El control de l'aigua a la Mallorca medieval* (Palma de Mallorca, El Tall, 1992), pp. 17–18, 34.

121 Reis Fontanals, 'Gestió d'un sistema hidràulic i interferències des d'un poder exterior: el cas de la Síquia de la Vila (Ciutat de Mallorca) als segles XIII i XIV', in press. In my article, 'Cap a una història institucional dels regs: un mètode d'estudi comparatiu', *Taller d'Història*, 3:1 (1994), 39–46, I was overly optimistic in evaluating the canal's return to regular operation.

122 Pierre Bonnassie, *From Slavery to Feudalism in South-Western Europe* (Cambridge, Cambridge University Press), p. 260.

the valley bordering a 'new canal' (*cequia nova*) and the river.[123] It is unclear whether Christian expansion of a previously existing Muslim system here resulted in less water for Muslim right-holders.

Most feudal lords, however, adopted an indirect strategy of allowing towns of autonomous communities of irrigators to go about their business unimpeded, thus sparing everyone concerned the high level of conflict that feudal intromission into the operating procedures of irrigation canals inevitably caused, then reaping the benefits of increased production through the normal avenues of seigniorial enrichment.[124] This is the conclusion of Furió and Martínez Sanmartín, in a study of the Horta del Cent, a vast irrigated space between Alzira and Xàtiva, comprising some twenty *alquerías* watered by the canals of Algirós and Enova. Here, in spite of the political fragmentation of the zone into a number of seigniorial jurisdictions, the irrigators of the two canals apportioned their own *séquiatge*, similar to the procedures adopted by the self-governing canals of the Valencian huerta.[125]

Feudalisation of Andalusi mills

I noted in Chapter 4 that what little we know of Andalusi mills, particularly their distribution, comes from mentions of them in the books of *Repartiment/Repartimiento*. There are several related issues in the assimilation of these mills by the Christian conquerors and new settlers. First is the social organisation of milling: in Al-Andalus, except for a relatively small number of mills owned by important individuals, mainly state officials, mills were a business enterprise like any other in or near cities, while in the *alquerías* they provided a service and were typically owned collectively by up to a dozen partners. Second is the political control of milling. In Christian Spain, and feudal Europe generally, mills were a regalian or seigniorial monopoly. At a 'banal' mill, peasants were obliged to grind their grain there and in no other place.[126] Many mills were not banal, but still constituted an important source of seigniorial income. Moreover, inasmuch as feudal or seigniorial rents were customarily

123 Ferran Garcia Garcia (= Garcia Oliver), *El naixement del monestir cistercenc de la Valldigna* (Valencia, Universitat, 1983), pp. 31–2.

124 See Chapter 5, above: 'Irrigation and Feudalism' (pp. 161–4).

125 Antoni Furió and Luis Pablo Martínez Sanmartín, 'De la hidràulica andalusi a la feudal: Continuitat i ruptura. L'Horta del Cent en l'Alzira medieval', *VI Assemblea d'Història de la Ribera* (Alzira, 1993), in press. In the sixteenth and seventeenth centuries, nevertheless, some lords attempted to extend their irrigated areas, violating the principle of proportionality.

126 As in Ballestar, Vall de Benifassa (Castellón); Sergi Selma Castell, 'Conquesta feudal i creació de monopolis de renda al País Valencià', *Boletín de la Sociedad Castellonense de Cultura*, 69 (1993), 333–55, on p. 335 n. 7.

payable in kind, and wheat flour was the preferred medium of payment, Muslims living under Christian rule had to devote more agricultural space to cereal cultivation than had been the practice before. In the *Repartiment* of Valencia, grants of mill and oven monopolies were systematically linked in order to structure demand for peasant taxes in a way that emphasised the production of wheat.[127] Third, there is the issue of the emplacement of mills within a hydraulic system. Barceló has argued that Andalusi mills were sited in a way that diminished conflict with irrigators whose priority of use of water was established. What then happened, first, when Andalusi hydraulic systems – including mills – were taken over by a society with different objectives and priorities in water use; and, second, if new mills were added to pre-existing hydraulic systems, how were they assimilated?

In Mallorca, pre-existing mills were fully feudalised, as at the *alquería* of Benicuaroz where the mill, granted to the same magnate as held the *alquería*, owed dues to the King and various other feudal lords as well.[128] A major feudal lord, Nunyo Sanç, alone held forty-two mills.[129] New mills were added to the hydraulic systems by the Catalans, frequently in inappropriate places, such as one in Coanegra built in the middle of a homogenous 'block' of irrigated parcels, rather than at the end, causing the loss of irrigable land.[130] In Valencia King James continued his policy of tightly controlling mills, granting some *alquerías* intact, but retaining the mills and ovens for himself,[131] or in other cases including the mill in the grant.[132] The King received dues of one-third to one-half of the profit of the mills he granted out.[133] James appointed a special collector for regalian mills and closely regulated the building of new ones.[134]

127 Selma Castell, 'Conquesta feudal', p. 336. In cursory review of later *Repartimientos*, I was only able to identify one more case of a grant linking milling and baking monopolies, in González, *Repartimiento de Sevilla*, II, 17, a grant to the Queen of Castile. Alfonso de Molino was granted an *aceña* (presumably a vertical grist-mill) with five stones, plus an indeterminate number of ovens (*hornos*) in Córdoba; Cabrera, 'Estructuras agrarias', p. 178 n. 21.

128 Helena Kirchner, 'La construcció de l'espai pagès: Les valls de Bunyola, Orient, Coanegra i Alaró a Mayurqa' (unpublished doctoral dissertation, Universitat Autònoma de Barcelona, Bellaterra, 1993), pp. 65–7.

129 *Remembrança de Nunyo Sanç*, eds. Mut Calafell and Rosselló Bordoy, pp. 130–4, 138.

130 Kirchner, 'Espai pagès', p. 111.

131 Cabanes Pecourt and Ferrer Navarro, eds., *Repartiment de Valencia*, I, 57 (nos. 0306, 0307).

132 *Ibid.*, p. 65, no. 407.

133 Sergi Selma Castell, 'Notes sobre la formació d'uns primers monopolis feudals a la Vall d'Albaida', *Alba* (Ontinyent), 7 (1992), 35–8, on p. 37.

134 Burns, *Medieval Colonialism*, pp. 54–5.

In Seville, Alfonso X gave all the mills of the city to the town council, nine of which were functioning and five of which had fallen down. They were given with the stipulation that water had to flow into the palaces of the city as it had in the time in the Moors.[135] In Carmona, also conquered in the thirteenth century, the Military Order of Calatrava was granted the 'dam of the mills of Remollena'.[136]

Vincent has found documentation for 237 mills in Granada in the early sixteenth century and estimates that the total must have been around 3,000. Many of these were abandoned between 1568 and 1593, perhaps as many as 80 per cent.[137] In the case of mills ruined in the fierce battle for Málaga, it was the norm *not* to grant licences to rebuild them on their former sites, even though the rule was not heeded in many cases.[138]

In the post-expulsion settlements of the 1570s, there was a standard norm established by the Crown. Mills were to be leased out, with the lessee responsible for their repair. Thus, in Fondón (Almería) settlers were to have the mills without payment for six years, with the obligation of repairing them.[139]

In Mudéjar Granada of the late fifteenth century, the collective ownership of mills was continued. Thus in Mondéjar, the ownership of the mill was apportioned among nine persons, the King and eight local Muslims – an unusual adaptation of the principle of seigniorial monopoly to local customs.[140] In Callosa d'en Sarriá, which is a typical mountain mesosystem fed from a rather large spring, Muslim shareholders (*parçoners*) still controlled the mill in the fifteenth century, even though it also had a feudal lord.[141] Similarly, in sixteenth-century Granada, rural grain mills were not seigniorial monopolies but were owned by groups of Moriscos. This was true even in the most seigniorialised areas: thus, in the *señorío* of the Enríquez, the lords controlled the ovens and stores, the Moriscos the mills. Unlike the prediction that milling would take on increased significance if taxes had to be paid in wheat, in the

135 González, *Repartimiento de Sevilla*, II, 322.

136 Manuel González Jiménez, 'Repartimiento de Carmona: Estudio y edición', *Historia, Instituciones, Documentos*, 7 (1980), 59–84, on p. 82.

137 Vincent, 'Société chrétienne almeriense', p. cv.

138 López de Coca Castañer, *Tierra de Málaga*, p. 120. But the contrary principle is asserted by Francisco Bejarano-Robles, 'El Repartimiento de Málaga. Introducción a su estudio', *Al-Andalus*, 31 (1966), 1–46, on p. 41: here he says that mills were to be re-established on their former sites, each with one *aranzada* of land.

139 Barrios Aguilera and Burriel Salcedo, *Repoblación de Granada*, p. 195. Ponce Molina, *Espacio agrario de Fondón*, p. 42.

140 Galán Sánchez, *Mudéjares del Reino de Granada*, p. 202.

141 Arxiu del Regne de València, Gobernació, Plets, 2271, 6th hand, fols. 1r–v, 47r–48v (10 January 1444).

Sierra de Filabres, mills had lower priority than before because the population was much lower.[142]

Continuity and change in post-*Repartimiento* settlement

Although I am in general agreement with the new doctrine of feudalism which holds that the level of peasant dependence in Spain was about on a par with that of countries of the feudal heartland,[143] I must also agree with Pierre Guichard and Manuel González Jiménez that the *Libros de Repartimiento* do not reflect it. They are not feudal documents. As I remarked above, they overwhelmingly document topographic rather than seigniorial space. Only a few thirteenth-century *Repartimientos* approach the quality and tenor of classical feudal land registers. In Mallorca, the *Rembrança* of Nunyo Sanç, with its detailed list of the appurtenances of Nunyo's fief, records feudal space. Likewise, in the *Repartimiento* of Seville (and no other in Andalucía, according to González Jiménez), there are a few true feudal grants, such as that of Borgabenzoar to the Bishop of Seville in return for an obligation to provide armed men.[144]

Overly zealous avatars of the new doctrine of Spanish feudalism see the malevolent tentacles of feudal aggression spreading everywhere. Yet the *Repartimientos* are decidedly non-feudal both in character and objectives.[145] In Valencia, Guichard notes the large segment of free property that the kings of Aragón sought to implant in the form of *hereditates* granted free and quit of any royal or seigniorial dues. If that freedom weakened towards the end of the thirteenth century with the steady progression of seigniorialism, that does not obviate the original objectives of the *Repartiment* to implant free townsmen and peasants.[146] Likewise, González Jiménez,

142 García Latorre, 'Pervivencia de los espacios agrarios', pp. 299–300, 305. See also Manuel Espinar Moreno, 'Noticias y materiales para el estudio del lugar de Alcázar en el Marquesado del Cenete (de la Edad Media a la expulsión de los Moriscos)', in *Homenaje a Prof. Darío Cabanelas Rodríguez* (Granada, Universidad, 1987), pp. 283–96, on p. 293, where the Molino Sierra had four Morisco owners. According to Ponce Molina, *Espacio agrario de Fondón*, p. 71, sixteenth-century mills were not seigniorialised, but some were muncipalised.

143 Thomas F. Glick, *Cristianos y musulmanes en la España medieval* (Madrid, Alianza, 1991), pp. 277–82, reflecting recent research on feudalism in Spain. Compare the views expressed in my *Islamic and Christian Spain in the Early Middle Ages: Comparative Perspectives on Social and Cultural Formation* (Princeton, Princeton University Press, 1970), pp. 210–14, which I no longer accept.

144 González Jiménez, 'Repartimientos andaluces del siglo XIII', pp. 96f.

145 Again, if one compares the *Repartimientos* as a whole with the *Domesday Book*, there can be no doubt of this assertion.

146 Pierre Guichard, 'Quelques notes à propos du repeuplement de Valence', in *Actas del Coloquio de la V Asamblea General de la Sociedad Española de Estudios Medievales* (Zaragoza, Diputación General de Aragón, 1991), pp. 121–34, on p. 131.

with respect to Andalucía, criticises those who posit reflexively that conquest led necessarily to seigniorialisation. The overwhelming mass of settlers were free men, free from archaic feudal services, with no substantial noble inroads until the end of the thirteenth century.[147] Thus in both Valencia and Andalucía, we can perceive a gradient of feudalisation, as the original objectives of the *Repartimientos* were submerged, against the original intent of royal policy, by the rising tide of seigniorialism, by the harshness of frontier life, and by a voraciously active land market that did not favour the small, rural freeholder. For these reasons the *Libros de Repartimiento* were not properly agromanagerial instruments because their social control function was nil, nor were they primarily taxation records.

The changing model of rural settlement that the *Llibres de Repartiment/Libros de Repartimiento* encodes reflects a number of discrete processes. First it is obvious that peoples of different cultures organise space differently because they perceive the environment differently – in the case before us, the *same* environment was perceived differently by Muslims and Christians. These perceptions in turn are based on idealised concepts of land-use related to very real systems of peasant production as well as to the differing ways in which rulers tax or otherwise relieve the peasantry of its 'surplus' production. The modal change must involve, as well, a selective paring down of the received register of Islamic culture and the reshaping of what was left according to the dictates of a different productive system. In terms of landscape change, including the institutions of irrigation and water allocation, I agree with García Latorre, Barrios Aguilera and others who argue that the reorganisation was 'soft' and much of the Islamic landscape was left in place. We can conclude both that the way in which Muslims had organised agricultural production struck the Christians as both productive and efficient and the elites were in general possessed of enough perspicacity that they did not wish to kill the goose that laid the golden egg. It is significant that the agrarian landscape is one aspect of culture that can be quickly purged of all of its ideological content. Mosques were changed to churches, and what remained

147 González Jiménez, 'Repartimientos andaluces' (1990), p. 117; González Jiménez, 'Conquista y repoblación de Andalucía. Estado de la cuestión cuarenta años después de la reunión de Jaca', in *Actas del Coloquio de la V Asamblea General de la Sociedad Española de Estudios Medievales* (Zaragoza, 1991), pp. 245–6, and *La repoblación de la zona de Sevilla durante el siglo XIV*, 2nd ed. (Seville, Universidad, 1993), pp. 92–5. Juan Torres Fontes makes the same point for Lorca – that royal policy was to establish a stable population of free peasants; *Repartimiento de Lorca*, 2nd ed. (Murcia, Academia Alfonso X el Sabio, 1994), p. lix.

was that part of Muslim economic and social organisation that could be accepted as worthwhile. Changes were made, nevertheless: a new model had to be devised for the administration of irrigation systems formerly run tribally; some accommodation, small or great, had to be made in the balance of crops to keep the feudal tribute system supplied with grain; this in turn meant that priorities in water use had to be skewed towards milling.

Manuel Barrios Aguilera identified an important mechanism, standard in all medieval and early modern *Repartimientos*, which acted in an automatic fashion to preserve the morphology of the Muslim landscape intact. But we must ask to what extent the actual organisation of the daily lives of Christian peasants cultivating the inherited Muslim landscape represents an assimilation of the practices, techniques or even customs of Muslim peasants. In situations where the two populations overlapped, clearly there was adequate opportunity for non-formal cultural diffusion to occur. By non-formal I mean any kind of cultural interchange not directed by formal governmental or paragovernmental institutions. Even in such cases, however, formal, directed culture change was probably more significant, as to a greater or lesser extent it was the policy of Christian authorities to keep the Muslim agricultural and hydraulic machines going. The extreme opposite end of the spectrum would include those villages where the entire Muslim population had fled before the new Christian arrived. There the new settlers found a perfect 'neutron bomb' scenario: houses, fields, canals, trees, and vines were intact: all that was needed was to get to work. Here we need recall López de Coca Castañer's admonition: this was in no way a wild and untamed frontier. Every aspect of the settlers' lives was directed by the authorities.[148] It is this dynamic that ensured, through formal mechanisms, the imposition of habits of social organisation (I have in mind the deliberate reinstitution of Morisco irrigation regimes in places where no settler could have had direct knowledge of them), and that the tenuous link with the previous culture and society was maintained.

The *Repartimientos*, considered as a chronological series, reflect changing land-tenure patterns in Al-Andalus which have yet to be analysed. They also record a tenurial revolution of major proportions where Christians replaced Muslims, although the impact of the latter upon the agrarian landscape (with respect to morphology of fields, parcellisation, and dispersion of parcels) was less disturbed and is still felt today.

148 López de Coca Castañer, *Tierra de Málaga*, p. 106.

7

Completing the transition

The completed transition was characterised by the diversification and enrichment of the register of material culture, institutions and ideas alike. The point can be made by examining three parallel phenomena: the diversification of pottery forms in late medieval Valencia as compared with the roster of forms available to early medieval Catalonia; the conceptual realignment of scientific disciplines in the wake of the translation movement of the twelfth and thirteenth centuries; and the patterning of certain urban institutions and conceptualisation of urban administration after Islamic progenitors in late medieval Valencia.

In comparative study, the most satisfactory analytical targets are phenomena which, first, have many variables, but also in which the variables divide conceptually into two or more sets. The greater the conceptual spread, the more useful the analysis. That is why pottery has been so successful an analytical tool in archaeology. Its variables relate either to form or to decoration, both registers displaying huge numbers of variants which lend themselves to systematic classification capable of generating powerful heuristic models. Form tells us about the function of the vessel, and function is easily linked to production and trade. Decoration relates to style and, among other uses, it provides sets of markers (both by itself and when conjoined with form) of great utility in cultural and social analysis.

So it is with science, the multiplication of whose disciplines in the Middle Ages was symptomatic of a general reorganisation of knowledge on the formal level. The content of individual disciplines provides a wealth of detail, particularly of a technical nature, that can itself be analysed for its cultural genealogy. Thus, the uses of astronomical lore forced a clear-cut division between astronomy and astrology, while the actual values contained, for example, in celestial tables, provide a different order of comparative data, one divorced from any overt ideology.

In the realm of institutions, I am most concerned with the set of urban administrative practices that the Arabs called *ḥisba*. The magistrate who carried out its precepts, the *muḥtasib*, was a combined market and public health inspector. Here too are a plethora of variables to analyse, in the multiplicity of ordinances regarding the various trades and categories of urban life that fell within the office's jurisdiction. But the nature of the jurisdiction can also be analysed, quite apart from its specific content, and such analysis relates to how urban life was conceptualised, not to the details of daily life.

Pottery: enriching the register

Rosselló Bordoy has noted that the number of different Arabic terms describing Andalusi pots rises from the eleventh to the thirteenth centuries, then turns down in the fourteenth and fifteenth. 'Does the semantic richness reflect morphological richness and variety of production in different stages of Andalusi daily life?' he asks. The rising curve suggests typological diversification, but the imprecision or polysemism of the Arabic terminology make it difficult to match each word with a specific archaeological artefact.[1] Literary sources for the names of Andalusi pots are particularly rich. The discovery of two eleventh-century notarial formularies, one of the Valencia *faqīh* al-Bunṭī, the other by the Toledan ibn Mughith, have clarified the terminology of medieval Andalusi ceramics considerably.[2] It is possible now to construct a kind of 'natural' classification (so to speak) of Andalusi pottery, using the real names of forms, rather than generic terms like *olla, marmita,* etc., which lack cultural precision.[3] The formularies of al-Bunṭī and ibn Mughith specify the volume of specific forms. That is, the State determined specific measures and these, in turn, influenced the daily life of the users of those particular pieces. The notion that some forms were subject to standardisation also provides another possible cultural marker, inasmuch as there was wide regional variation in

1 Guillermo Rosselló Bordoy, 'Precisiones sobre terminología cerámica andalusí', in *Coloquio Hispano-Italiano de Arqueología Medieval* (Granada, Patronato de la Alhambra, 1992), pp. 253–62, on pp. 254–5; and 'Las cerámicas de primera época: Algunas observaciones metodológicas', in Antonio Malpica Cuello, ed., *La cerámica altomedieval en el sur de Al-Andalus* (Granada, Universidad, 1993), pp. 13–25, on pp. 19, 21.

2 Pablo Yzquierdo y Pino, 'Fonts documentals i ceràmica andalusina. Alguns suggeriments', *Segones Jornades de Joves Historiadors i Historiadores* (Barcelona, mimeo, 1988), n.p.

3 This point has been a *cri de guerre* of Rosselló Bordoy. See his *El nombre de las cosas en Al-Andalus: Una propuesta de terminología cerámica* (Palma de Mallorca, Museu de Mallorca, 1991).

standard measures.[4] It may be, therefore, that the variety of ceramic forms was growing just at the moment when Christians were coming into contact with forms of everyday Muslim life by virtue of acquiring large enclaves of *mudéjars* after the conquest of Toledo in 1085.

The contrast between the pottery register in early medieval Catalonia and post-conquest Valencia, settled mainly by Catalan peasants and bourgeois, is particularly instructive.[5] Ceramics had played a relatively minor role in Catalonia of the twelfth century and earlier. Plates and other common kitchenwares used by the peasantry were frequently made of wood, some wooden utensils, like the *escudella*, giving their names to ceramic copies. Catalan potteries, where the well-studied grey-wares were produced, were primitive compared to those not only of the Islamic world but even to Castile and León. Early medieval Catalan pottery was unglazed and had scant decoration. When the Catalans conquered Valencia, however, they found a technically sophisticated ceramic infrastructure already in place, with a numerous Muslim peasantry used to ceramic rather than wooden vessels. There, a fusion of Muslim and Christian ceramic traditions took place, with striking results: the number of forms made by Christians increased from fourteen (mainly variants on a basic, all-purpose cooking pot – *olla*) in the pre-conquest Catalan repertory to forty-three thereafter. An earlier *escudella* (bowl) was replaced by one influenced by the Arab *jofaina*. Those forms most closely associated with specific eating habits, such as *ollas* and *cazuelas*, had the clearest Catalan precedents, although they were now supplied with an interior lead glaze, an innovation of Muslim inspiration. The rest of the kitchenwares were from the Islamic repertory and most were now known in Catalan by a heavily Arabised terminology (e.g., *alfàbia, ancolla, gerra, setra, cassola,* and so forth). Christian potters learned how to use more complex ceramic techniques, such as the use of oxidising glazes and faster wheel-throwing, together with a more complex form of production in more efficient kilns located in towns rather than in rural areas, as had been the case in Catalonia. The small Catalan kiln of Roman inspiration yielded – in Valencia – to the typical Persian-style updraught kiln of the great ceramic centres of the Islamic world.[6]

4 Yzquierdo y Pino, 'Fonts documentals'.

5 This section is based on Jaume Coll i Conesa, 'Ceràmica i canvi cultural a la València medieval. L'impacte de la conquesta', *Afers*, 7 (1988–89), 125–67; and Jaume Coll i Conesa, Javier Martí Oltra and Josefa Pascuela Pacheco, *Cerámica y cambio cultural* (Valencia, Ministerio de Cultura, 1988).

6 Compare the Roman-style Catalan kilns described by Manuel Riu Riu, 'Talleres y hornos de alfareros de cerámica gris en Cataluña', in *Fours de potiers et 'testares'*

Science: expanding the quadrivium

The received division of arts and sciences known throughout early medieval Europe was the familiar arrangement of the seven liberal arts of Roman learning: the trivium (broadly, the humanities) of grammar, rhetoric, and logic, and the quadrivium (broadly the sciences), which included arithmetic, geometry, astronomy, and music. In the course of the first three Islamic centuries so much new information and so many new techniques (most importantly those of Indian astronomy and of positional arithmetic) were ingested that science as actually practised no longer resembled the classical quadrivium. It was thus that al-Fārābī (*c.* 870–950), a Persian philosopher working in ᶜAbbāsid Bagdad, devised a new classification of the sciences in an influential book entitled *Iḥṣa' al-ᶜulūm*, the catalogue or enumeration of the sciences. In al-Fārābī's classification mathematics (still including music) now consisted of seven, rather than four, sciences: arithmetic, geometry, optics, astronomy, music, weights and measures, and engineering (a broadly inclusive category that included algebra, scientific instruments, architecture, and other practical applications). Physics, which had no real place in the old quadrivium, al-Fārābī reckoned as having seven subdivisions.[7] Al-Fārābī's new classification was disseminated in Spain and Latin Europe generally by a canon named Domingo González (better known by his Latin name, Dominicus Gundisalvus or Gundisalinus), whose treatise *De divisione philosophiae* was an interpretation of the partial text of al-Fārābī to which Gundisalvus had access. The Arabs' approach to astronomy was highly practical: its heart was the use of instruments like the astrolabe to determine the position of celestial bodies and, on the basis of observation, to determine latitude with the aid of astronomical tables. The new emphasis on observation required the development of a new theoretical superstructure, which in turn required the remodelling of Aristotelian and Ptolemaic doctrine. In any event, al-Fārābī and his Latin followers made a distinction, based on this experiential context, between practical and theoretical sciences which in part transcended the disciplinary array itself.

The style of science that was stressed in Latin Europe was in

médiévaux en Méditerranée occidentale (Madrid, Casa de Velázquez, 1990), pp. 105–15 (drawing on p. 107), with the Valencian 'Arab kiln' described by Josep A. Gisbert Santonja, 'Los hornos del alfar islámico de la Avda. Montgó/Calle Teulada, casco urbano de Denia (Alicante),' in *ibid.*, pp. 75–91 (drawing on p. 91).

7 Al-Fārābī, *Catálogo de las ciencias*, trans. Angel González Palencia, 2nd ed. (Madrid, CSIC, 1953), pp. xv (mathematics), 60–2 (physics).

part determined by the style practised in Al-Andalus. In Beaujouan's description:

The lack of interest in abstract mathematics, the predominance of astronomy and astrology in the early translations, the relatively late date of the Arabo-Latin versions of Aristotle's natural philosophy, the failure to use important works by eastern Arabic scholars: all are explained by the evolution of Arabic science in the Iberian peninsula, with its peculiarities of history and geography, its particularist pride within the Islamic world, its conditioning by the oppressive domination of the Malikite faqihs.[8]

In arithmetic there were, first of all, competing systems of calculation, both subsequent to the introduction, also by the Andalusis, of positional calculation. In one system, the abacus was used, with counters in nine columns, but without the zero. In the other, not adopted in Latin Europe until the twelfth century, operations were performed in sand or dust (the antecedent of chalk and slate), with a zero: this method was called *algorismus* in Latin, after the author of the text which diffused it, al-Khwārizmī. But the method associated with al-Khwārizmī was not only conceptual (positional, with zero) but technical: dust-writing made possible the opportunity to make successive corrections by erasing, thus decreasing dependence on memory.[9] There was, moreover, no agreement as to what the basic operations of arithmetic were, nor in what order they belonged.[10]

There was also a related change in medieval encyclopedias, owing as well to the influence of Arabic upon Latin thought. As a result of the translation of Arabic and Greek scientific treatises into Latin, emphasis shifted from factual information organised by terminological criteria (as in the *Etymologies* of Isidore of Seville, or in the medieval bestiaries) to an organisational norm based on the nature of the individual sciences and systems and principles of knowledge. This process led to a redefinition of principles and the generation or reception of new sciences unknown to the ancients. A multiplicity of epistemological approaches made it possible to consider a phenomenon from a variety of angles: the laws of motion

8 Guy Beaujouan, 'The Transformation of the Quadrivium', in Robert L. Benson and Giles Constable, eds., *Renaissance and Renewal in the Twelfth Century* (Cambridge, MA, Harvard University Press, 1982), pp. 463–87, on p. 465. See also Thomas F. Glick, 'Science in Medieval Spain: The Jewish Contribution in the Context of *Convivencia*', in Vivian Mann, Thomas F. Glick and Jererilynn Dodds, eds., *Convivencia: Jews, Muslims and Christians in Medieval Spain* (New York, George Braziller, 1992), pp. 83–111, on pp. 99–100.

9 Beaujouan, 'Transformation of the Quadrivium', pp. 467–9.

10 See examples in Glick, 'Science in Medieval Spain', p. 98.

could be sought through induction from observed motions, or by dialectical methods, and so forth.[11] Thus McKeon concludes that in the case of the movement of scientific and philosophical translation the influence of Arab culture on Latin was not so much in the diffusion of discrete elements, but in 'the rearrangement of schemata which they shared and the modification of the data, methods, and truths organised in those schemata to new specifications and evidence'.[12]

In the transition from Late Antiquity to Early Islam, we noted that the Islamic polity was so radically distinct from those of antiquity that elements of classical civilisation were ingested in an inchoate way. What can we now say about Christian appropriation of the 'debris' of Al-Andalus? Beaujouan describes the relationship between early medieval Latin scholasticism and the new Arabic science to currents of water that don't mix well owing to different levels of salinity. It took a new mind-set which arose in twelfth-century universities and was diffused in new textbooks to finally make the currents mix (that is, to make the assimilation of new epistemologies possible).[13]

What this kind of analysis of the translation movement and its results shows is the highly selective nature of the processes involved. The Andalusis were selective in the kind of sciences they practised or emphasised: they did not exploit the entire corpus available to them. By the same token, the Christians who translated Arab science into Latin or Romance were more interested in some parts of the corpus than in others. As one edged closer to matters philosophical, moreover, it was possible to borrow the methods of Arabic science without necessarily embracing its ideological content.

The translation movement of the twelfth and thirteenth centuries, in my view, constituted a kind of mental repopulation, a preparation for the colonisation by Christians of a foreign mental universe: the superimposition of Christian upon Muslim mental structures (although typically portrayed as the reverse).

Urban institutions: adopting an Islamic model

The Islamic *muḥtasib* was a typical product of the compounding of institutions out of the debris of antiquity. The Muslims found two

11 Richard McKeon, 'The Organization of Sciences and the Relations of Cultures in the Twelfth and Thirteenth Centuries', in J. Murdoch and E. Scylla, eds., *The Cultural Context of Medieval Learning* (Dordrecht, D. Reidel, 1975), pp. 151–92, on pp. 184, 186–7.

12 *Ibid.*, p. 157.

13 Beaujouan, 'Transformation of the Quadrivium', p. 484.

Greek urban magistrates in the conquered cities of the Byzantine empire: the *agoranomos*, a market inspector, and the *astynomos*, who supervised the streets and public spaces of the city. These they conjoined into an all-round urban magistracy appointed by the chief religious judge, or *qāḍi*, of the city. Like the *qāḍi* there could by only one to a city (this is called 'unipersonality'), and he could act summarily to punish routine infringements of market, building, and public health codes. The magistracy was found useful by the Spanish Christians who adopted it in the thirteenth century as the *almotacen* in Castile, the *mustassaf* in Valencia and Catalonia.[14]

As I have noted, the *muḥtasib* diffused to Christian Spain along with a grab-bag of ordinances and regulations pertaining to specific crafts and to various other aspects of town life (such as privacy-protecting regulations), an amalgam of customary, mainly secular, practices that typify the vague 'seepage' of Islamic institutions that pervaded Christian society of the later Middle Ages, difficult to pin down owing to its general nature.

In general terms, the *mustassaf* was an important official in towns of the Crown of Aragón, particularly in the kingdom of Valencia and, to a lesser degree, the other Catalan-speaking regions.[15] If anything, the *mustassaf* was *more* important and significant an official in Valencia than his immediate antecedent had been in Al-Andalus.[16] Why this should have been so is a complex question, but there is no question that the *mustassafia* jurisdiction was particularly congruent with the kind of legal system found in Valencia and in the Crown of Aragón generally. In the first place, Christian Spain provided a much more receptive climate for the development of an area of administration based on customary law than was possible in the Islamic world. That is because *sharī ʿa* applies universally and no specific allowance is made for specialised jurisdictions. Second, the office was provided for in the *Furs* of Valencia, the legal code promulgated by James I, which gave it a firm legal standing. Then, as the *Furs* were 'territorialised', that same jurisdiction was automatically reproduced in dozens of urban locales. Third, as the jurisprudential traditions of Roman law, with its rigorous courtroom

14 For a general perspective, see Thomas F. Glick, 'Muḥtasib and Mustasaf: A Case Study of Institutional Diffusion', *Viator*, 2 (1971), 59–81, and Pedro Chalmeta, *El señor del zoco en España* (Madrid, Instituto Hispano-Arabe de Cultura, 1973).

15 The following paragraphs are based on my article, 'New Perspectives on the *Ḥisba* and its Hispanic Derivatives', *Al-Qanṭara*, 13 (1992), pp. 475–89, on pp. 481, 487–8.

16 At its inception, the *mustassafia* was borrowed from a strong living model: al-Saqaṭī, *muḥtasib* of Málaga and author of a *ḥisba* manual, was a contemporary of James I who named the first Valencian *mustassaf*.

practices and evidentiary procedures, gained force, there was an immediate benefit to the legal system as a whole in granting broad latitude to the efficient and economical (both in cost and in time) summary procedures of special, customary jurisdictions, such as the urban jurisdiction represented by the *mustasaffia*, or, to be sure, the vast expanse of customary irrigation law.

Although the Muslims attempted to invest the hybrid *agoranomos/astynomos* that was the *muḥtasib* with a moral or religious rationale in keeping with the spirit of the *sharīʿa*, the religious basis of the magistracy was in large part a myth. The enjoinder to encourage the good (*al-amr b'il-maʿrūf*) is nothing more than conventional piety, and all religions define their civic programmes in similar terms. Andalusi *muḥtasib*/s, according to Ibn Saʿīd, attempted to codify this body of law (in the 'Manuals of *ḥisba*') 'as if it were jurisprudence' (that is, *fiqh*),[17] when in reality it was merely a conglomeration of customary, mainly secular, practices. Ultimately the power of the *muḥtasib* depended not upon any religious sanction but rather upon the will of the state and the norms of administration adopted.

From Christian documentation we learn about the diffusion of Islamic institutions which, although transformed, bear within them certain organising principles (e.g., combination of the agoranomic and astynomic, unipersonality, summary justice, orality,[18] lack of functional differentiation) which shed light both on Christian practice and on the antecedent Islamic one. The two jurisdictions were similar but the societies in which they were embedded were radically different.

By comparing manuals of *ḥisba* and of *mustassafia* – in reality, compilations of practical ordinances regulating craft guilds and public conduct – we know which of these elements was transmitted from Al-Andalus to Christian Spain. Pedro Chalmeta collated Al-Saqaṭī's thirteenth-century treatise from Málaga with sixteenth-century municipal ordinances from Málaga, Seville, and Granada. Among the close similarities noted were such police functions as the manner of punishing infractions in the market; the specific manner in which the baker of unlawful bread is to be punished (loaves broken into small pieces); the responsibility of shopkeepers for the conduct of their employees; administrative matters such as the specifications for the manufacture of weights; and details regarding purveyors of specific provisions: how fine flour is to be prepared; that a side of meat may not be swelled up by artificial

17 Cited by Chalmeta, *El señor del zoco*, p. 312.

18 The *mustassaf*, like the *muḥtasib*, acted 'sumariament i de paraula' in Valencian legal parlance; Vicent Josep Escartí, 'La documentació escrita generada pel mustassaf', *Saitabi* (Valencia), 41 (1991), 189–99, on p. 190.

means to increase its weight; that different cuts of meat and different kinds of fish had to be weighed separately. These administrative traditions were set in the practice of market-places in Christian Andalusia by the end of the Middle Ages.[19] But there is more beneath the surface as well – unarticulated, unrecognised, and unconscious – which has to do with the rationale underlying *ḥisba* jurisdiction. That jurisdiction was only partially defined by its subject matter. Unipersonality lent further definition: it channelled practice in a certain way. In the medieval Islamic cities, all officials had the same jurisdictional area and convention alone determined what applied to the *ḥisba*, the *shurṭa* (police), etc. All magistracies dependent on the *qaḍā'* (office of *qāḍi*) had administrative, judicial *and* executive functions.

We might well ask why the *ḥisba* jurisdiction showed greater vitality in Valencia than in Al-Andalus, or in the Islamic world generally where, towards the end of the Middle Ages, the magistracy entered a long decline. As a locally-based institution in an urban Islamic culture lacking its own administrative machinery, it bore from its inception the potential for conflict with the central power of the State.[20] It became inevitably implicated in the collection of non-canonical taxes. For this reason, Muslim traditionalists today who call for a return to 'Islamic' government, have not favoured the resurgence of the *muḥtasib*. Nor did nationalists speak for it, because of its locally-based nature.[21] In the 'garrison state' of early Islam there was much broader latitude for what was at root a secular magistracy than would be true in the pervasively Islamic states of the lower Middle Ages and more modern times.[22] For all these reasons, the office was successful in Christian Spain, where the foral tradition favoured local jurisdictions, and the summary jurisdiction of the *mustassaf* was viewed as a welcome alternative to complicated court procedures, rather than as a skirting of canonical law.

Archaeology and cultural change in medieval Spain

The new medieval archaeology, in a symbiotic relationship with the historiography of *incastellamento*, has generated a series of interesting

19 Chalmeta, 'El "Kitāb fī adab al-ḥisba" (Libro del buen gobierno del zoco) de al-Saqaṭī', *Al-Andalus*, 32 (1967), 125–62, 359–97; 33 (1968), 143–95, 367–434, on pp. 376 n. 1, 377 n. 3; 379 n. 2, 394 n. 4 (vol. 32); 145 n. 4, 166 n. 2, 169 n. 2, 172 n. 1 (vol. 33).

20 See, on this score, G. E. von Grunebaum, Islam: *Essays in the Nature and Growth of a Cultural Tradition* (New York, Barnes and Noble, 1961), p. 149.

21 L. P. Harvey, personal communication, 11 March 1993.

22 Islamic law had a 'horror of exemptive privileges' (Jean Sauvaget, quoted by von Grunebaum, *Islam*, p. 152.).

questions and, by 'universalising' them, has forced medievalists to respond, thereby refining the models in question while at the same time confirming, to a substantial degree, the original hypotheses, establishing a new and potent paradigm. That paradigm is a general model of rural settlement in Al-Andalus that I have here called the *ḥiṣn/qarya* complex. It involves the organisation of rural settlement and peasant production in units consisting of a castle, a small number of villages, and irrigation agriculture. The major architects of the paradigm have been Guichard, Bazzana, and Cressier, with important contributions by Acién Almansa, Azuar Ruiz, M. Barceló, Sénac, Torró, and several others.

A constellation of subsidiary hypotheses have contributed to the expansion and enrichment of the basic paradigm. First is Guichard's proposal that *alquerías* established by Berbers (and by extension, Arabs) were tribally organised and representative of segmentary social forms. His evidence is highly eclectic and includes comparative data from North Africa, documentary evidence, and, above all, the interpretation of place-names in order to establish the relationship between tribalism and settlement. This is not in any way an arcane solution, because the high incidence of Beni- names in Al-Andalus is readily apparent. 'Extensive' archaeology suggested how the pieces of the puzzle could be put together. Evidence gathered by subsequent scholars has tended to confirm both the prevalence of Berber clans and the fact that tribes, whether Berber or Arab, were still segmenting at the time that Beni- names appeared, suggesting that such settlements were established throughout the history of Al-Andalus. Barceló and Poveda Sánchez have made important confirmatory contributions to Guichard's hypothesis, with participation by many scholars, each contributing the identification of a toponym or two.

Acién Almansa's interpretation of the social dynamics of the *fitna* of the tenth century, associating Ibn Ḥafṣūn and the Muwallad leadership with the last gasp of Gothic proto-feudalism as it survived, Arabised and Islamised, in the mountains of southern Spain, is the finest single piece of historical reinterpretation to emerge from the new archaeology. The key to unravelling the mystery was the logic of settlement and the differentiation among types of *ḥuṣūn* in the mountains of eastern Andalucía.

Barceló's ongoing research on design of irrigation systems is another hypothesis that emerged from the *ḥiṣn/qarya* paradigm and aims to demonstrate the ways in which segmentary social organisation dictated the organisation of irrigation-system layouts. This line of research, with important participation by Kirchner, Martí, Navarro, and Selma Castell, has changed the focus of the study of irrigation layouts from technology to social organisation, a direct result of the influence of the new paradigm. The institutional pieces of this puzzle are in large part still missing; I have tried to supply some of them here.

Examination of pottery associated with *ḥuṣūn* provided the immediate stimulus for a theorisation of the transition from Roman to Islamic Spain and the conceptualisation of a transitional 'Paleoandalusi' culture. Gutiérrez Lloret has been most influential in this area, with contributions by Zozaya and many others. The associated research in Paleoandalusi agriculture has involved Azuar Ruiz, Gutiérrez Lloret, and Mateu Bellés.

Finally, the impetus that the *incastellamento* debate has given to the reconceptualisation of feudalism has meant that a number of archaeologist/historians working on Al-Andalus have perforce been drawn into the process of reconceptualising the nature of feudalism in Christian Spain. The sharp feudal/non-feudal dichotomy that informs the comments of Guichard, Barceló, Martí, and others on Catalan and Valencian feudalism has been a necessary step in defining the context in which the reconceptualisation would take place, thereby drawing the feudalism debate in eastern Spain into the orbit of the Italian discussion, paralleling the close contacts between those who have studied *incastellamento* in Al-Andalus and Italian medieval archaeologists, in particular those associated with the journal *Archeologia Medievale.*

Theorising cultural change

Here the history of irrigation can supply examples illustrating the broader problem of how cultural change in medieval Spain might be theorised. In the case of long-term, large-scale cultural change initiated by conquest, there was bound to be an appropriation of techniques already on the ground, which were integrated slowly into a cultural and social system radically different than the one in which they had been established. From the perspective of technology, Rodríguez López and Cara Barrionuevo sum up the problem well:

> Technological traditions become confused with one another and overlap. Small innovations succeed one another in social practice either from constant use or from the interpretation of traces that remain in the landscape. But if it is impossible to deny the existence of a pre-Islamic irrigation agriculture and its ample technological baggage, the conditions for

their generalisation (= diffusion) are social and can only be realised in a particular historical moment.[23]

We cannot always look for a *particular* set of institutions and techniques that can be ascribed to one specific settlement group. Berber/Arab migration can be characterised in the same way that Foster characterises the passage of Spanish institutions to the New World: the total cultural resources of the Arab Islamic world or even of the Magrib were far greater than the peninsula could absorb: 'The potential cultural mixture was too rich.' Conquerors had to form a new society in some kind of symbiotic relationship with the Hispano-Roman population. 'But there were limits to the complexity of these new societies and communities, particularly in a frontier epoch. In terms of functional analysis there is an optimum number (or level of development) of institutions and traits.' Thus, the post-conquest population could not have possibly consumed *all* the Berber and Arab influences to which it was exposed and combine them with *all* the traits of Hispano-Roman society.[24]

What does the highly documented second transition tell us about the virtually undocumented first one? First, that conquerors move into existing landholds when they can. Second, that they practice agriculture that is familiar to them. Third, that they exploit new niches consonant with their style of agriculture and organisation of peasant labour. Fourth, they may take advantage of any improvements (irrigation canals, fencing, buildings) that they find on the ground, but will later make them fit their own cultural or cultivating style. Fifth, the differential organisation of peasant labour as the primary motor of landscape change and continuity – unless formally imposed by a governmental institution (improbable in the earlier transition) – is more apparent than real.

23 Juana M. Rodríguez López and Lorenzo Cara Barrionuevo, 'Aproximación al conocimiento de la historia agrícola de la Alpujarra oriental (Almería). Epocas antigua y medieval', *AZA*, I, 441–66, on p. 459. They go on, then, to make an unwarranted assumption: that the generalisation of irrigation systems in the Alpujarras was only possible after the successful conclusion to the pacification programme of ʿAbd al-Raḥmān III, as if social peace were a prerequisite of tribally-based irrigation.

24 Following and paraphrasing George Foster, *Culture and Conquest* (Chicago, Quadrangle, 1960), p. 14.

Select bibliography

Abbreviations

AZA *El agua en zonas áridas: Arqueología e historia.* I Coloquio de Historia y Medio Físico. 2 vols. Almería, Instituto de Estudios Almerienses, 1989.
I CAME *I Congreso de Arqueología Medieval Española.* 4 vols. Zaragoza, 1986.
II CAME *II Congreso de Arqueología Medieval Española.* 3 vols. Madrid, 1987.
III CAME *III Congreso de Arqueología Medieval Española.* 2 vols. Oviedo, 1989.
IV CAME *IV Congreso de Arqueología Medieval Española.* 3 vols. Alicante, 1993.
Castrum 2 *Structures de l'habitat et occupation du sol dans les pays méditerranéens: Les méthodes et l'apport de l'archéologie extensive,* ed. Ghislaine Noyé. Madrid/Rome, Casa de Velázquez/Ecole Française de Rome, 1988.
Castrum 3 *Guerre, fortification et habitat dans le monde méditerraneén au Moyen Age.* Madrid/Rome, Casa de Velázquez/Ecole Française de Rome, 1988.
Castrum 4 *Frontière et peuplement dans le monde méditerranéen au Moyen Age.* Madrid/Rome, Casa de Velázquez/Ecole Française de Rome, 1992.

Books and articles

Acién Almansa, Manuel. *Ronda y su serranía en tiempo de los Reyes Católicos.* 3 vols. Málaga, Universidad, 1979.

—— 'Cerámica a torno lento en Bezmiliana. Cronología, tipos y difusión', *I CAME,* IV, 243–6.

—— 'Poblamiento y fortificación en el sur de Al-Andalus. La formación de un país de *ḥuṣūn*', *III CAME,* I, 135–50.

—— 'Recientes estudios sobre arqueología andalusí en el sur de Al-Andalus', *Aragón en la Edad Media,* 9 (1991), 355–69.

—— 'Arqueología medieval en Andalucía', in *Coloquio Hispano-Italiano de Arqueología Medieval* (Granada, Patronato de la Alhambra, 1992), pp. 27–33.

—— 'Sobre la función de los *ḥuṣūn* en el sur de Al-Andalus. La fortificación en el Califato', in *ibid.,* pp. 263–74.

—— 'La cultura material de época emiral en el sur de Al-Andalus. Nuevas perspectivas', in Antonio Malpica Cuello, ed., *La cerámica altomedieval en el sur de Al-Andalus* (Granada, Universidad, 1993), pp. 153–72.

—— *Entre el feudalismo y el Islam. ʿUmar Ibn Ḥafṣūn en los historiadores, en las fuentes y en la historia.* Jaén, Universidad, 1994.

Aguadé Nieto, Santiago. 'Molino hidráulico y sociedad en Cuenca durante la edad media (1177–1300)', *Anuario de Estudios Medievales,* 12 (1982), 241–77.

Alfonso Antón, Isabel. *La colonización cisterciense en la meseta del Duero. El dominio de Moreruela (siglos XII–XIV).* Zamora, Diputación Provincial, 1986.

Almería entre dos culturas. Siglos XIII al XVI. 2 vols. Almería, Instituto de Estudios Almerienses, 1990.

Alvarez Delgado, Yasmina. 'Cerámicas comunes con y sin decoración, siglo X. Arcávica (Cuenca)', *II CAME,* II, 403–12.

Amigues, François. 'Poitiers mudejares et chrétiens dans la région de Valence', *Archéologie Islamique,* 3 (1992), 129–67.

Amouric, Henri. 'De la roue horizontale à la roue verticale dans les moulins à eau. Une révolution technologique en Provence?', *Provence Historique,* 33 (1983), 157–69.

Andújar Castillo, Francisco. 'Adaptación y dominio del agua. La vega de Almería en el primer tercio del siglo XVII', *AZA,* II, 1087–99.

Araguas, Philippe, 'Le réseau castral en Catalogne vers 1350', *Castrum 3,* pp. 113–22.

Arce, Javier. *El último siglo de la España romana (284–409).* Madrid, Alianza, 1982.

—— 'La transformación de Hispania en época tardorromana: Paisaje urbano, paisaje rural', in *De la Antigüedad al Medioevo, siglos IV–VIII. III Congreso de Estudios Medievales* (Madrid, Fundación Sánchez-Albornoz, 1993), pp. 225–49.

Arié, Rachel. *L'Espagne musulmane au temps des Nasrides (1232–1492).* Paris, Brocard, 1973.

Azuar Ruiz, Rafael. 'Una interpretación del 'ḥiṣn' musulmán en el ámbito rural', *Revista de Estudios Alicantinos,* 37 (1982), 33–41.

—— *Denia islámica: Arqueología y poblamiento.* Alicante, Diputación Provincial, 1989.

—— 'La rábita califal de Guardamar y el paleoambiente del Bajo Segura (Alicante) en el siglo X', *Boletín de Arqueología Medieval,* 5 (1991), 135–50.

Azuar Ruiz, Rafael and Gutiérrez Lloret, Sonia. 'Formación y transformación de un espacio agrícola islámico en el sur del País Valenciano: El Bajo Segura (siglos IX–XIII)', *Castrum 5,* in press.

Banks, Philip. 'The Roman Inheritance and Topographical Transitions in Early Medieval Barcelona', in T. F. Blagg *et al.*, eds., *Papers in Iberian Archeology,* BAR International Series 193 (1984), part ii, pp. 552–77.

Barbero, Abilio and Vigil, Marcelo. *La formación del feudalismo en la Península Ibérica* (1978). 4th ed. Barcelona, Crítica, 1986.

Barceló, Miquel. 'Les plagues de llagost a la Carpetània', *Estudis d'Història Agraria,* 1 (1978), pp. 67–84.

—— 'La primerenca organització fiscal d'Al-Andalus segons la "Crònica del 754" (95/713–138/755)', *Faventia,* 1 (1979), 231–61.

—— *Sobre Mayurqa.* Palma de Mallorca, Quaderns de Ca La Gran Cristiana, 1984.

—— 'La qüestió de l'hidraulisme andalusí', in Barceló *et al.*, *Les aigües cercades,* pp. 9–36.

—— 'Aigua i assentaments andalusins entre Xerta i Amposta (s. VII–XII)', *II CAME,* 2:413–20.

—— 'Vísperas de feudales. La sociedad de Sharq al-Andalus justo antes de la conquista catalana', in Felipe Maíllo Salgado, ed., *España, Al-Andalus, Sefarad: Síntesis y nuevas perspectivas* (Salamanca, Universidad, 1988), pp. 99–112.

—— 'El diseño de espacios irrigados en Al-Andalus: Un enunciado de principios generales', *AZA,* I, xiii–l.

—— 'Assentaments berbers i àrabs a les regions del nord-est d'Al-Andalus: el cas de l'Alt Penedès (Barcelona)', in *La Marche Supérieure d'Al-Andalus et l'Occident Chrétien.* Madrid, Casa de Velázquez, 1991, pp. 89–97.

—— 'La cuestión septentrional. La arqueología de los asentamientos andalusíes más antiguos', *Aragón en la Edad Media,* 9 (1991), 341–53.

—— 'Quina arqueologia per al-Andalus', in *Coloquio Hispano-Italiano de Arqueología Medieval* (Granada, Patronato de la Alhambra, 1992), pp. 243–52.

—— 'Historia y arqueología', *Al-Qanṭara,* 13 (1992), 457–62.

—— '¿Por qué los historiadores académicos prefieren hablar de islamización en vez de hablar de campesinos?', *Archeologia Medievale,* 19 (1992), 63–73.

Barceló, Miquel (ed.). *Arqueología medieval: En las afueras del 'medievalismo'.* Barcelona, Crítica, 1988.

Barceló, Miquel, Pinyol, Joan, and Poveda Sánchez, Angel. 'Eren ramaders els rafals de Mayurqa? Un exercici de simulació històrica', in *Les Illes Orientals d'Al-Andalus.* Palma de Mallorca, Institut d'Estudios Baleàrics, 1987, pp. 115–122.

Barceló, Miquel *et al. Les aigües cercades (Els qanat(s) de l'illa de Mallorca).* Palma de Mallorca, Institut d'Estudis Baleàrics, 1986.

—— 'Arqueología: La Font Antiga de Crevillent: Ensayo de descripción arqueológica', *Areas. Revista de Ciencias Sociales* (Murcia), 9 (1988), 217–31.

Barceló Torres, Carmen. *Toponimia arábica del País Valencià. Alqueries i castells.* Valencia, Diputación Provincial, 1983.

—— '¿Galgos o podencos? Sobre la supuesta berberización del País Valenciano en los siglos VIII y IX', *Al-Qanṭara,* 11 (1990), 429–60.

Bardhan, Pranab. 'Symposium on Management of Local Commons', *Journal of Economic Perspectives,* 7 (1993), 87–92.

Barral i Altet, Xavier. 'Quelques exemples d'habitat groupé en hauteur en Catalogne (Xe–XIe siécles)', *Castrum 2,* pp. 85–96.

Barrios Aguilera, Manuel. *Libro de los Repartimientos de Loja.* Granada, Diputación Provincial, 1988.

—— 'La repoblación del Reino de Granada por Felipe II', in Miguel Angel Ladero Quesada, ed., *La incorporación de Granada a la Corona de Castilla* (Granada, Diputación Provincial, 1993), 607–28.

—— *Moriscos y repoblación en las postrimeras de la Granada islámica.* Granada, Diputación, 1993.

Barrios Aguilera, Manuel and Burriel Salcedo, Margarita M. *La repoblación de Granada después de la expulsión de los Moriscos.* Granada, Universidad, 1986.

Barrios García, Angel. 'Toponomástica e historia: Notas sobre la despoblación en la zona meridional del Duero', in *Estudios en memoria del Profesor D. Salvador de Moxó* (Madrid, Universidad, 1982), I, 115–34.

—— *Estructuras agrarias y de poder en Castilla. El ejemplo de Avila (1085–1320).* 2 vols. Salamanca, Universidad, 1983.

—— 'Repoblación de la zona meridional del Duero. Fases de ocupación, procedencias y distribució espacial de los grupos repobladores', *Studia Historica (Historia Medieval),* 3:2 (1985), 33–82.

Bazzana, André. 'Typologie . . . : Les habitats fortifiés du Sharq al-Andalus', in *Habitats fortifiés et organisation de l'espace en Méditerranée médiévale* (Lyon, Maison de l'Orient, 1983), pp. 19–27.

—— 'Les structures: fortification et habitat', in *ibid.*, pp. 161–72.

—— *Maisons d'Al-Andalus: Habitat médiéval et structures du peuplement dans l'Espagne orientale.* 2 vols. Madrid, Casa de Velázquez, 1992.

Bazzana, André, Climent, Salvador, and Montmessin, Yves. *El yacimiento medieval de 'Les Jovades'—Oliva (Valencia).* Gandía, Ayuntamiento, 1987.

Bazzana, André, Cressier, Patrice and Guichard, Pierre. *Les châteaux ruraux d'Al-Andalus.* Madrid, Casa de Velázquez, 1988.

Bazzana, André *et al.* 'L'Hydraulique agraire dans l'Espagne médiévale', in *L'Eau et les hommes en Méditerranée,* ed. André de Reparaz. Marseilles, CNRS, 1987, pp. 43–66.

Bazzana, André and Guichard, Pierre. 'La frontière du Sharq al-Andalus', in *La Marche Supérieure d'Al-Andalus et l'Occident Chrétien* (Casa de Velázquez, Universidad de Zaragoza, 1991), pp. 77–8.

—— 'Irrigation et société dans l'Espagne orientale au Moyen Age', in *L'homme et l'eau en Mediterranée et Proche Orient,* ed. J. Metral and P. Sanlaville (Lyon, Maison de l'Orient, 1981), pp. 115–40.

Bazzana, André, Guichard, Pierre, and Segura Martí, José María. 'Du *ḥiṣn* musulman au *Castrum* chrétien: La chateau de Perpunchent (Lorcha, province d'Alicante)', *Mélanges de la Casa de Velázquez,* 18 (1982), 449–65.

Beaujouan, Guy. 'The Transformation of the Quadrivium', in Robert L. Benson and Giles Constable, eds., *Renaissance and Renewal in the Twelfth Century* (Cambridge, MA, Harvard University Press, 1982), pp. 463–87.

Bejarano-Robles, Francisco. 'El Repartimiento de Málaga. Introducción a su estudio', *Al-Andalus,* 31 (1966), 1–46.

Bejarano-Robles, Francisco and Vallvé, Joaquín. *Repartimiento de Comares (1487–1496).* Barcelona, Universidad, 1974.

Berque, Jacques. *Structures sociales du Haut-Atlas.* Paris, Presses Universitaires de France, 1955.

Bertrand, Maryelle, and Cressier, Patrice. 'Irrigation et aménagement du terroir dans la vallé de l'Andarax (Almería): Les réseaux anciens de Ragol', *Mélanges de la Casa de Velázquez,* 21 (1985), 115–35.

Bevià, Màrius. 'L'Albacar musulmà del castell d'Alacant', *Sharq al-Andalus,* 1 (1984), 131–40.

Bois, Guy. *The Transformation of the Year One Thousand.* Manchester, Manchester University Press, 1992.

Bolós i Mascalans, Jordi, and Fàbregas i Sabater, Miquel. 'Els molins de la conca mitjana del Llobregat durant l'alta edat mitjana. I. Introducció', *Quaderns d'Estudis Medievals,* 3 (1982), 556–8.

Bolós i Mascalans, Jordi, and Martínez i Huelde, Angel. 'El molí de la Torre Baldovina de Santa Coloma de Gamenet (Barcelonès)', *Acta Historica et Archaeologica Mediaevalia,* 7–8 (1986–87), 421–35.

Bolós i Mascalans, Jordi, and Nuet i Badia, Josep. *Els molins fariners.* Barcelona, Ketres, 1983.

Bolós i Mascalans, Jordi, and Padilla Lapuente, Iñaki. 'Un molí d'origen medieval: El Molinet de Naval', *Quaderns d'Estudis Medievals,* 1 (1980), 49–55.

Bonnassie, Pierre. *La Catalogne du milieu du Xe à la fin du XIe siècle: Croissance et mutations d'une société.* 2 vols. Toulouse, Université de Toulouse-Le Mirail, 1975.

—— *From Slavery to Feudalism in South-Western Europe.* Cambridge, Cambridge University Press, 1991.

Bonnassie, Pierre, *et al. Estructuras feudales y feudalismo en el mundo mediterráneo.* Barcelona, Crítica, 1984.

Borrero Fernández, Mercedes. 'Las transformaciones de la estructura de la propiedad de la tierra en la Baja Andalucía en la segunda mitad del siglo XIII', in *Andalucía entre Oriente y Occidente (1236–1492)* (Córdoba, Diputación Provincial, 1988), pp. 191–208.

Bresc, Henri. 'Féodalité coloniale en terre d'Islam: La Sicile (1070–1240)', in *Structures féodales et féodalisme dans l'Occident méditerranéen (Xe–XIIe siècles): Bilan et perspectives de recherches* (Rome, Ecole Française de Rome, 1980), pp. 631–47.

—— 'Les eaux siciliennes: une domestication inachevée du XIIe au XVe siècle', in Elisabeth Crouzet-Pavan and Jean-Claude Maire-Vigueur, eds., *Water Control in Western Europe, Twelfth–Sixteenth Centuries* (Milan, Università Bocconi, 1994; Proceedings, Eleventh International Economic History Congress, vol. B2), 73–85.

Bulliet, Richard. *The Camel and the Wheel.* Cambridge, MA, Harvard University Press, 1975.

—— *Conversion to Islam in the Medieval Period: An Essay in Quantitative History.* Cambridge, MA, Harvard University Press, 1979.

—— 'Conversion to Islam and the Emergence of a Muslim Society in Iran', in Nehemia Levtzion, ed., *Conversion to Islam* (New York, Holmes and Meier, 1979), pp. 38–51.

Burns, Robert I. *Medieval Colonialism: Postcrusade Exploitation of Islamic Valencia.* Princeton, Princeton University Press, 1975.

—— *Muslims, Christians, and Jews in the Crusader Kingdom of Valencia.* Cambridge, Cambridge University Press, 1984.

—— *Society and Documentation in Medieval Valencia.* Princeton, Princeton University Press, 1985.

—— *Foundations of Crusader Valencia* (*Diplomatarium,* II). Princeton, Princeton University Press, 1991.

Busquets Mulet, Jaime. 'El códice latinoarábigo del Repartimiento de Mallorca (parte latina)', *Butlletí de la Societat Arqueològica Lul·liana,* 30 (1947–52), 6–55.

—— 'El códice latinoarábigo del Repartimiento de Mallorca (texto árabe)', in *Homenaje a Millás Vallicrosa,* 2 vols. (Barcelona, CSIC, 1954–56), I, 243–300.

Butzer, Karl W. 'Castles on the Valencian Border March', *Al-ᶜUsur al-Wusta: Bulletin of Middle East Medievalists,* 4, 2 (October 1992), 17–19, 39.

Butzer, Karl W. *et al.* 'Irrigation Agrosystems in Eastern Spain: Roman or Islamic Origins?', *Annals of the Association of American Geographers,* 75 (1985), 479–509.

—— 'Orígenes de la distribución intercomunitaria del agua en la Sierra de Espadán (País Valenciano)', in *Los paisajes del agua* (Valencia/Alicante, Universidad, 1989), pp. 223–8.

Caballero Zoreda, L. 'Perviviencia de elementos visigodos en la transición al mundo medieval. Planteamiento del tema', *III CAME,* I, 111–34.

Cabanes Pecourt, María Desamparados, and Ferrer Navarro, Ramón, eds. *Libre del Repartiment del Regne de Valencia.* 3 vols. Zaragoza, Textos Medievales, 1979.

Cabrera, Emilio. 'Evolución de las estructuras agrarias en Andalucía a raíz de su reconquista y repoblación', in *Andalucía entre Oriente y Occidente (1236–1492)* (Córdoba, Diputación Provincial, 1988), pp. 171–89.

Cabrillana Cizar, Nicolás. *Moriscos y cristianos en Yunquera.* Málaga, n.p., 1994.
Carbajo Serrano, María José. *El monasterio de los santos Cosme e Damian de Abellar.* León, Centro de Estudios e Investigación San Isidro, 1988.
Carbonero Gamundi, María Antonia. 'Terrasses per al cultiu irrigat i distribucío social de l'aigua a Banyalbufar (Mallorca)', *Documents d'Anàlisi Geogràfica,* 4 (1983), 32–68.
Carmona González, Pilar. 'Interpretación paleohidrológica y geoarqueológica del substrato romano y musulmán de la ciudad de Valencia', *Saitabi,* 40 (1990), 163–76.
Casa hispano-musulmana, La. Aportaciones de la arqueología. Granada, Patronato de la Alhambra, 1990.
Casado Alonso, Hilario. *Señores, mercaderes y campesinos. La comarca de Burgos a fines de la edad media.* n.p., Junta de Castilla y León, 1987.
Castillo Galdeano, Francisco and Martínez Madrid, Rafael. 'Producciones cerámicas en Bajjana', in Antonio Malpica Cuello, ed., *La cerámica altomedieval en el sur de Al-Andalus* (Granada, Universidad, 1993), pp. 67–117.
Castro, Américo. *The Spaniards.* Berkeley, University of California Press, 1971.
Catalán, Diego, and Andrés, María Soledad de, eds. *Crónica del Moro Rasis.* Madrid, Gredos, 1974.
Cerrillo Martín de Cáceres, E., and Fernández Corrales, José María. 'Contribución al estudio del asentamiento romano en Extremadura. Análisis espacial aplicado al S. de Trujillo', *Norba,* 1 (1980), 157–75.
Chabás, Roque. *Distribución de las aguas en 1244 y donaciones del término de Gandía por D. Jaime I.* Valencia, Francisco Vives Mora, 1898.
Chalmeta, Pedro. *El señor del zoco en España.* Madrid, Instituto Hispano-Arabe de Cultura, 1973.
Clariana, Joan Francesc, and Prevosti, Marta. 'Sobre la pervivencia de hábitats rurales romanos en la Alta Edad Media en el Maresme', *II CAME,* III, 429–36.
Coll i Conesa, Jaume. 'Ceràmica i canvi cultural a la València medieval. L'impacte de la conquesta', *Afers,* 7 (1988–89), 125–67.
Coll i Conesa, Jaume, Martí Oltra, Javier, and Pascuela Pacheco, Josefa. *Cerámica y cambio cultural.* Valencia, Ministerio de Cultura, 1988.
Coloquio Hispano-Italiano de Arqueología Medieval. Granada, Patronato de la Alhambra, 1992.
Coope, Jessica A. 'Religious and Cultural Conversion to Islam in Ninth-Century Umayyad Cordoba', *Journal of World History,* 4 (1993), 47–68.
—— *The Martyrs of Cordoba: Community and Family Conflict in an Age of Mass Conversion.* Lincoln, University of Nebraska Press, 1995.
Corriente, Federico. *Arabe andalusí y lenguas romances.* Madrid, Mapfre, 1992.
Cressier, Patrice. 'Archéologie des structures hydrauliques en al-Andalus', *AZA,* I, li–xcii.
—— 'Agua, fortificaciones y poblamiento: El aporte de la arqueología a los estudios sobre el sureste peninsular', *Aragón en la Edad Media,* 9 (1991), 403–27.
—— *Estudios de arqueología medieval en Almería.* Almería, Instituto de Estudios Almerienses, 1992.
Crone, Patricia. *Slaves on Horses: The Evolution of the Islamic Polity.* Cambridge, Cambridge University Press, 1980.
—— *Roman, Provincial and Islamic Law: The Origins of the Islamic Patronate.* Cambridge, Cambridge University Press, 1987.
Crone, Patricia, and Cook, Michael. *Hagarism: the Making of the Islamic World.* Cambridge, Cambridge University Press, 1977.
Curchin, Leonard A. '*Vici* and *Pagi* in Roman Spain', *Revue des Etudes Anciennes,* 87 (1985), 327–43.
—— *The Local Magistrates of Roman Spain.* Toronto, University of Toronto Press, 1990.
Dallière-Benelhadj, Valérie. 'Le "Château" en Al-Andalus: un problème de terminologie', in *Habitats fortifiées et organisation de l'espace en Méditerranée médiévale* (Lyon, Maison de l'Orient, 1983), pp. 63–7.

De Al-Andalus a la sociedad feudal. Barcelona, CSIC, 1990.
Debord, André. 'The Castellan Revolution and the Peace of God in Aquitaine', in Richard Landes and Thomas Head, eds., *The Peace of God: Social Violence and Religious Response in France around the Year 1000* (Ithaca, Cornell University Press, 1992), pp. 135–64.
Delaigue, Marie-Christine. 'Mutations de l'espace villageois en Andalousie orientale. Effets immédiats et lointains de la Reconquête', *Mélanges de la Casa de Velázquez,* 26 (1990), 131–62.
Díaz García, Amador and Barrios Aguilera, Manuel. *De toponimia granadina.* Granada, Universidad, 1991.
Domínguez García, Manuel. 'La acequia de riegos de Motril y las ordenanzas de 1561', *AZA,* II, 951–68.
Epalza, Míkel de. 'Funciones ganaderas de los albacares, en las fortalezas musulmanas', *Sharq al-Andalus,* 1 (1984), 47–54.
—— 'La islamización de Al-Andalus. Mozárabes y neomozárabes', *Revista del Instituto Egipcio de Estudios Islámicos en Madrid,* 23 (1985–86), 171–9.
—— 'Relacions dels països catalans amb el món musulmà', *Revista de Catalunya* (February 1987), pp. 49–62.
—— 'Mozarabs: An Emblematic Christian Minority in Islamic Al-Andalus', in Salma K. Jayyusi, ed., *The Legacy of Muslim Spain* (Leiden, E. J. Brill, 1992), pp. 149–70.
Epalza, Míkel de, and Llobregat, Enrique. '¿Hubo mozárabes en tierras valencianas? Proceso de islamización del levante de la península (Sharq al-Andalus)', *Revista de Estudios Alicantinos,* 36 (1982), 7–31.
Epalza, Míkel, and Rubiera, María Jesús. 'La *sofra* (*sujra*) en el Sharq al-Andalus (s. XIII)', *Sharq al-Andalus,* 3 (1986), 33–8.
—— 'Estat actual dels estudis de toponimia valenciana d'origen àrab', *Xé Col·loqui General de la Societat d'Onomàstica. Ier d'Onomàstica Valenciana* (Valencia, 1986), pp. 420–6.
Escacena Carrasaco, José Luis. 'Yacimientos arqueológicos de la época medieval en el flanco oriental del Aljarafe', *II CAME,* II, 579–87.
Escartí, Vicent Josep. 'La documentació escrita generada pel mustassaf', *Saitabi* (Valencia), 41 (1991), 189–99.
Español Bertran, Francesc. 'Els casals de molins medievals a les comarques tarragonines. Contribució a l'estudi de la seva tipologia arquitectònica', *Acta Historica et Archaeologica Mediaevalia,* 1 (1980), 231–54.
Espinar Moreno, Manuel. 'Los árboles en las Tierras de Cantoria. Suertes confeccionadas y reparto', *Roel,* 6 (1985), 139–69.
—— 'Reparto de las aguas del Río Abrucena (1273?–1420)', *Revista del Centro de Estudios Históricos de Granada y su Reino*', 2nd epoch, 1 (1987), 69–94.
—— 'Noticias y materiales para el estudio del lugar de Alcázar en el Marquesado del Cenete (de la Edad Media a la expulsión de los Moriscos)', in *Homeraje a Prof. Darío Cabanelas Rodríquez.* Granada, Universidad, 1987, pp. 283–96.
—— 'Estudio sobre propiedad particular de las aguas de la acequia de Jarales (1267–1528). Problemas de abastecimeinto urbano y regadíos de tierras entre las alquerías de Abrucena y Abla', *AZA,* I, 247–66.
—— 'Población y agricultura de una alquería almeriense en los siglos XII y XIII', in *Almería entre dos culturas. Siglos XVIII al XVI,* 2 vols. (Almería, Instituto de Estudios Almerienses, 1990), 187–207.
—— 'Bizar: Una alquería musulmana y el paso al dominio cristiano (siglos XII–XVI)', in *Andalucía entre Oriente y Occidente (1236–1492)* (Córdoba, Diputación Provincial, 1988), pp. 707–18.
Espinar Moreno, Manuel, Quesada Gómez, Juan José, and Amezcua Pretel, José. 'Materiales romanos, visigodos y árabes en la autovia de circunvalación de Granada. Aportaciones a la arqueología y cultura material', in *In Memoriam J. Cabrera Moreno* (Granada, Universidad, 1992), pp. 103–16.
Espinar Moreno, Manuel, Glick, Thomas F., and Martínez Ruiz, Juan. 'El término árabe

dawla "turno de riego", en una alquería de las tahas de Berja y Dalias: Ambroz (Almería)', *AZA*, I, 121–41.

Espinar Moreno, Manuel, and Quesada Gómez, María Dolores. 'El regadío en el distrito del castillo de Sant Aflay. Repartimiento del Río de la Regua (1304–1524)', *Estudios de Historia y Arqueología Medievales* (Cádiz), 5–6 (1985–86), 127–57.

Estepa Díaz, Carlos. 'In Memoriam: Abilio Barbero de Aguilera (1931–1990)', *En la España Medieval*, 14 (1991), 11–17.

Faci, Javier. 'Vocablos referentes al sector agrario en León y Castilla durante la Alta Edad Media', *Moneda y Crédito*, 144 (March 1978), 69–87.

Farabi, al-. *Catálogo de las ciencias*. Trans. Angel González Palencia. 2nd ed. Madrid, CSIC, 1953.

Fentress, Elizabeth. 'Forever Berber?', *Opus* (Rome), 2 (1983), 161–75.

Fernández Corrales, José M. 'El asentamiento rural romano en torno a los cursos alto y medio del Salor: su marco geográfico y distribución', *Norba*, 4 (1983), 207–21.

Fité i Llevot, Francesc. 'Un apropament a l'estudi dels molins del Montsec i la Vall d'Ager', *Acta Historica et Archaeologica Mediaevalia*, 4 (1983), 207–38.

Fleming, Robin. *Kings and Lords in Conquest England*. Cambridge, Cambridge University Press, 1991.

Formació i expansió del feudalisme català. Estudi General (Girona), 5–6 (1985–86).

Fossier, Robert. *Enfance de l'Europe*. 2 vols. Paris, Presses Universitaires de France, 1982.

Furió, Antoni and Garcia, Ferran. 'Dificultats agràries en la formació i consolidació del feudalisme al País Valencià', *Estudi General*, 5–6 (1985–86), 291–310.

Furió, Antonio and Martínez Sanmartín, Luis Pablo. 'De la hidràulica andalusí a la feudal: Continuïtat i ruptura. L'Horta del Cent en l'Alzira medieval', VI Assemblea d'Història de la Ribera (Alzira, 1993), in press.

—— 'Assuts i molins sobre el Xúquer en la Baixa Edat Mitjana', *IV CAME*, III, 575–86.

Galán Sánchez, Angel. *Los mudéjares del Reino de Granada*. Granada, Universidad, 1991.

García de Cortázar, José Angel. *El dominio del monasterio de San Millán de la Cogolla (siglos X a XII)*. Salamanca, Universidad, 1969.

—— 'El equipamiento molinar en la Rioja Alta en los siglos X al XIII', *Studia Silensia*, 3 (1976), 387–405.

—— 'Del Cantábrico al Duero', in J. A. García de Cortázar, ed., *Organización social del espacio en la España medieval* (Barcelona, Ariel, 1985), pp. 43–83.

—— 'La repoblación del Valle del Duero en el siglo IX: Del yermo estratégico a la organización social del espacio', in *La reconquista y repoblación de los reinos hispánicos: Estado de la cuestión de los últimos cuarenta años* (Zaragoza, Diputación General de Aragon, 1991), pp. 15–40.

García de Cortázar, José Angel, and Díez Herrera, Carmen. *La formación de la sociedad hispano-cristiana del Cantábrico al Ebro en los siglos VIII a XI*. Santander, Estudio, 1982.

García de Valdeavellano, Luis. *Sobre los burgos y los burgueses de la España medieval*. Madrid, Real Academia de la Historia, 1960.

García Latorre, Juan. 'La pervivencia de los espacios agrarios y los sistemas hidráulicos de tradición andalusí tras la expulsión de los moriscos', *Revista del Centro de Estudios Históricos de Granada y su Reino*, 2nd epoch, 6 (1992), 297–317.

—— 'Arqueología medieval e historia moderna en el reino de Granada. El caso de la Sierra de Filabres', *Chronica Nova*, 20 (1992), 177–207.

García Moreno, Luis A. *Historia de la España visigoda*. Barcelona, Cátedra, 1989.

García Oliver, Ferran. *El naixement del monestir cistercenc de la Valldigna*. Valencia, Universitat, 1983.

—— *Terra de feudals*. Valencia, Institució Valenciana d'Estudis i Investigacio, 1991.

García Tapia, Nicolás and Carricajo Carbajo, Carlos. *Molinos de la provincia de Valladolid*. Valladolid, Cámara Oficial de Comercio, 1990.

Gautier-Dalché, Jean. 'Moulin à eau, seigneurie, communauté rurale dans le nord de l'Espagne (IXe–XIIe siècles)', in *Etudes de Civilisation Médiévale. Mélanges offerts à Edmond-René Labande* (Poitiers, CESCM, 1974), pp. 337–49.

—— 'Châteaux et peuplements dans la Péninsule Ibérique (Xe–XIIIe siècles)', in

Châteaux et peuplements en Europe Occidentale du Xe au XVIIIe siècle (Auch, Centre Cultural de l'Abbaye de Flaran, 1980), pp. 93–107.

—— 'Reconquête et structures de l'habitat en Castille', *Castrum* 3, 199–206.

Gil Albarracín, Antonio. 'Los regadíos del bajo Andarax durante el siglo XVI', *AZA*, II, 969–80.

Gisbert Santonja, Josep A. 'Los hornos del alfar islámico de la Avda. Montgó/Calle Teulada, casco urbano de Denia (Alicante)', in *Fours de potiers et 'testares' médiévaux en Méditerranée occidentale* (Madrid, Casa de Velázquez, 1990), pp. 75–91.

Glick, Thomas F. *Irrigation and Society in Medieval Valencia.* Cambridge, MA, Harvard University Press, 1970.

—— 'Muḥtasib and Mustasaf: A Case Study of Institutional Diffusion', *Viator*, 2 (1971), 59–81.

—— *Islamic and Christian Spain in the Early Middle Ages: Comparative Perspectives on Social and Cultural Formation.* Princeton, Princeton University Press, 1979.

—— 'Las técnicas hidráulicas antes y después de la conquista', in *En torno al 750 aniversario: Antecedentes y consecuencias de la conquista de Valencia*, 2 vols. (Valencia, Consell Valencià de Cultura, 1989), I, 53–71.

—— 'Molins d'aigua a l'Horta medieval de València', *Afers* (Valencia), 9 (1990), 9–22.

—— 'El sentido arqueológico de las instituciones hidráulicas. Regadío berber y regadío español', *Aragón vive su historia. II Jornadas de Cultura Islámica* (Madrid, Instituto Occidental de Cultura Islámica, 1990), pp. 165–71.

—— 'Regadío y técnicas hidráulicas en al-Andalus. Su difusión según un eje Este-Oeste', *La caña de azúcar en tiempos de los grandes descubrimientos, 1450–1550* (Motril, Casa de la Palma, 1990), pp. 83–98.

—— 'Sir Clements Markham i l'interés britànic en el regadiu hispànic a mitjan segle XIX', in C. R. Markham, *Informe sobre el regadiu de l'Espanya de l'Est (1867)* (Valencia, Edicions Alfons el Magnànim, 1991), pp. 7–44.

—— *Cristianos y musulmanes en la España medieval.* Madrid, Alianza, 1991.

—— 'Historia del regadío y las técnicas hidráulicas en la España medieval y moderna. Bibliografía comentada', *Chronica Nova* (Granada), 18 (1990), 121–53; 19 (1991), 167–92; 20 (1992), 209–32.

—— 'New Perspectives on the *Ḥisba* and its Hispanic Derivatives', *Al-Qanṭara*, 13 (1992), 475–89.

—— *Tecnología, ciencia y cultura en la España medieval.* Madrid, Alianza, 1992.

—— 'Science in Medieval Spain: The Jewish Contribution in the Context of *Convivencia*', in Vivian Mann, Thomas F. Glick and Jererilynn Dodds, eds., *Convivencia: Jews, Muslism and Christians in Medieval Spain* (New York, George Braziller, 1992), pp. 83–111.

—— 'Irrigació en l'Horta de València durant el segle XV', in *Lluís de Santangel: Un nou home, un nou món* (Valencia, Generalitat Valenciana, 1992), pp. 147–54.

—— 'Hydraulic Technology in Al-Andalus', in Salma K. Jayyusi, ed., *The Legacy of Muslim Spain* (Leiden, E. J. Brill, 1992), pp. 974–86.

—— 'Sobre la tipologia convencional dels molins hidràulics', *Afers*, 15 (1993), 53–6.

—— 'La alta edat mitjana', in *Història del País Valencià* (Valencia, Tres i Quatre, 1992), pp. 57–82.

—— 'Cap a una història institucional dels regs: un mètode d'estudi comparatiu', *Taller d'Història*, 3:1 (1994), 39–46.

—— 'Arthur Maass y el análisis institucional del regadío en España', *Arbor*, 151 (1995), 13–33.

—— 'Berbers in Valencia: the Case of Irrigation', in *Medieval Spain and the Western Mediterranean: Essays in Honor of Robert I. Burns, S. J.* (Leiden, E. J. Brill, 1995).

Gómez Becerra, Antonio. *El Maraute (Motril): Un asentamiento medieval en la costa de Granada.* Motril, Ayuntamiento, 1992.

—— 'Cerámica a torneta rodente de "El Maraute" (Motril). Una primera aproximación a la cerámica altomedieval de la costa granadina', in Antonio Malpica Cuello, ed., *La cerámica altomedieval en el sur de Al-Andalus* (Granada, Universidad, 1993), pp. 172–91.

Gómez Cruz, Manuel. 'Las ordenanzas de riego de Almería año 1755', *AZA*, II, 1101–26.

González, Julio. *Repartimiento de Sevilla.* 2 vols. Madrid, CSIC, 1951.

—— *Repoblación de Castilla la Nueva.* 2 vols. Madrid, Universidad Complutense, 1975–76.

González Blanco, Antonio. 'La población del sureste durante los siglos oscuros (IV–X)', *Antigüedad y cristianismo* (Murcia), 5 (1988), 11–27.

González Jiménez, Manuel. 'Repartimiento de Carmona. Estudio y edición', *Historia, Instituciones, Documentos,* 7 (1980), 59–84.

—— *En torno a los orígenes de Andalucía.* 2nd ed. Seville, Universidad, 1988.

—— 'Repartimientos andaluces del siglo XIII: Perspectiva de conjunto y problemas', in *De Al-Andalus a la Sociedad feudal: Los repartimientos bajomedievales* (Barcelona, CSIC, 1990), pp. 95–117.

—— 'Conquista y repoblación de Andalucía. Estado de la cuestión cuarenta años después de la reunión de Jaca', in *La reconquista y repoblación de los reinos hispánicos. Actas del Coloquio de la V Asamblea General de la Sociedad Española de Estudios Medievales* (Zaragoza, Diputación General de Aragón, 1991), pp. 233–48.

—— *La repoblación de la zona de Sevilla durante el siglo XIV.* 2nd ed. Seville, Universidad, 1993.

González Jiménez, Manuel, and González Gómez, Antonio. *El Libro de Repartimiento de Jerez de la Frontera.* Cádiz, Instituto de Estudios Gaditanos, 1980.

Goody, Jack. *The Development of the Family and Marriage in Europe.* Cambridge, Cambridge University Press, 1983.

Gorges, Jean-Gerard. *Les villas Hispano-Romaines.* Paris, E. de Broccard, 1979.

—— 'Implantations rurales et réseau routier en zone meritaine: Convergences et divergences', *Caesarodunum,* 18 (1983), 413–24.

—— 'Prospections archéologiques autour d'Emerita Augusta: Soixante-dix sites ruraux en quête de signification', *Revue des Etudes Anciennes,* 88 (1986), 215–36.

Greene, Kevin. *The Archeology of the Roman Economy,* Berkeley, University of California Press, 1986.

Grima Cervantes, Juan A. *La expulsión morisca y la repoblación cristiana de Turre (1570–1596).* Almería, Diputación Provincial, 1988.

Guichard, Pierre. *Al-Andalus: Estructura antropológica de una sociedad islámica en Occidente.* Barcelona, Barral, 1976.

—— 'La Valencia musulmana: Los siglos oscuros', in *Nuestra Historia* (Valencia, Mas Ivars, 1980), II, 208–30.

—— 'Del califato a la conquista cristiana', *ibid.*, II, 237–62.

—— 'La conquista militar y la estructuración política del Reino', *ibid.*, III, 13–42.

—— 'La repoblación y la condición de los musulmanes', *ibid.*, III, 43–82.

—— 'Las transformaciones sociales y económicos', *ibid.*, III, 83–108.

—— 'El castillo y el valle de Pop durante la edad media: Contribución al estudio de los señoríos valencianos', *Anales de la Universidad de Alicante. Historia Medieval,* 2 (1983), 19–31.

—— 'El problema de la existencia de estructuras de tipo "feudal" en la sociedad de Al-Andalus', in P. Bonnassie *et al.*, *Estructuras feudales y feudalismo en el mundo mediterráneo* (Barcelona, Crítica, 1984), pp. 117–45.

—— 'Les Mozarabes de Valence et d'Al-Andalus entre l'histoire et le mythe', *Revue de l'Occident Musulman et de la Mediterranée,* 40 (1985), 17–27.

—— *Estudios sobre historia medieval.* Valencia, Diputación Provincial, 1987.

—— 'El impacto de la Reconquista en la sociedad musulmana', in *Historia del Pueblo Valenciano* (Valencia, Levante, 1988), pp. 221–40.

—— 'Le problème des structures agraires en Al-Andalus avant la conquête chrétienne', in *Andalucía entre Oriente y Occidente (1236–1492)* (Córdoba, Diputación Provincial, 1988), pp. 161–70.

—— 'A propos des raḥals de l'Espagne orientale', *Miscelanea Medieval Murciana,* 15 (1989), 11–24.

—— 'Faut-il en finir avec les berbers de Valencia?', *Al-Qanṭara,* 11 (1990), 461–73.

—— *Les Musulmans de Valence et la Reconquête (XIe–XIIIe siècles).* 2 vols. Damascus, Institut Français de Damas, 1990–91.

—— 'La toponymie tribale berbère valencienne: Réponse à quelques objections philologiques', in *Festgabe für Hans-Rudolph Singer* (Frankfurt, Peter Lang, 1991), pp. 125–41.

—— 'Quelques notes à propos du repeuplement de Valence', in *La reconquista y repoblación de los reinas hispánicos. Actas del Coloquio de la V Asamblea General de la Sociedad Española de Estudios Medievales* (Zaragoza, Diputación General de Aragón, 1991), pp. 121–34.

—— 'Els Berbers de València, una vegada més. Resposta a Carme Barceló', *Afers,* 15 (1993), 225–32.

Guichard, Pierre, and Bazzana, André. 'La sociedad musulmana valenciana en vísperas de la conquista cristiana', *Nuestra Historia* (Valencia, Mas Ivars, 1980), II, 263–80.

Gutiérrez Lloret, Sonia. *Cerámica común paleoandalusí del sur de Alicante (siglos VII–X).* Alicante, Caja de Ahorros Provincial, 1988.

—— 'El poblamiento tardorromano en Alicante a través de los testimonios materiales: Estado de la cuestión y perspectivas', *Antigüedad y cristianismo, V. Arte y poblamiento en el SE península durante los últimos siglos de civilización romana* (Murcia, Universidad, 1988), 323–37.

—— 'Espacio y poblamiento paleoandalusí en el sur de Alicante: Orígen y distribución', *III CAME,* II, 341–5.

—— 'De la civitas a la madīna: Destrucción y formación de la ciudad en el sureste de Al-Andalus. El debate arqueológico', *IV CAME,* I, 13–35.

—— 'La formación de Tudmir desde la periferia del estado islámico', *Cuadernos de Madīnat al Zahrā,* III (in press).

—— 'El tránsito de la antigüedad tardía al mundo islámico en la cora de Tudmir: Cultura material y poblamiento paleoandalusí'. Unpublished doctoral dissertation, University of Alicante, 1992.

—— 'La cerámica paleoandalusí del sureste península (Tudmir): Producción y distribución (siglos VII al X)', in Antonio Malpica Cuello, ed., *La cerámica altomedieval en el sur de Al-Andalus* (Granada, Universidad, 1993), pp. 37–65.

—— 'El origen de la huerta de Orihuela entre los siglos VII y XI: Una propuesta arqueológica sobre la explotación de las zonas húmedas del Bajo Segura', *Arbor,* 151, 1995, 65–93.

—— 'La producción de pan y aceite en ambientes domésticos. Límites y posibilidades de una aproximación etnoarqueológica', 'Formas de habitar e alimentação na Idade Media' (Mértola, 17–20 September 1993), typescript.

Hernández Benito, Pedro. *La vega de Granada a fines de la Edad Media según las rentas de los habices.* Granada, Diputación Provincial, 1990.

Hitchcock, Richard. 'Arabic Proper Names in the Becerro de Celanova', in *Cultures in Contact in Medieval Spain: Historical and Literary Essays Presented to L. P. Harvey* (London, King's College, 1990), pp. 111–26.

Holt, Richard. *The Mills of Medieval England.* Oxford, Basil Blackwell, 1988.

Illes orientals d'Al-Andalus, Les. Palma de Mallorca, Institut d'Estudis Baleàrics, 1987.

Isla Frez, Amancio. *La sociedad gallega en la alta edad media.* Madrid, CSIC, 1992.

Jarrega Domínguez, Ramón. 'Notas sobre una forma cerámica. Aportación al estudio de la transición del mundo romano al medieval en el este de Hispania', *I CAME,* II, 305–13.

Jayyusi, Salma K., ed. *The Legacy of Muslim Spain.* Leiden, E. J. Brill, 1992.

Jiménez Jurado, María Isabel. 'La ruralización de Almería en el siglo XVI. Problemas económicos derivados de la irrigación de la tierra', *AZA,* II, 1005–15.

Jular Pérez-Alfaro, Cristina. '*Alfoz y tierra* a través de documentación castellana y leonesa de 1157 a 1230. Contribución al estudio del *dominio señorial*', *Studia Histórica. Historia Medieval,* 9 (1991), 9–42.

Keay, Simon. 'Decline or Continuity? The Coastal Economy of the Conventus Tarraconensis from the Fourth Century until the Late Sixth Century', in T. F. Blagg *et al.*, eds., *Papers in Iberian Archeology,* BAR International Series 193 (1984), pt. ii, 552–77.

—— *Roman Spain.* Berkeley, University of California Press, 1988.

Kirchner, Helena. 'La construcció de l'espai pagès: Les valls de Bunyola, Orient, Coanegra i Alaró a Mayurqa.' Unpublished doctoral dissertation, Universitat Autònoma de Barcelona, Bellaterra, 1993.

—— 'Construir el agua. Irrigación y trabajo campesino en la edad media', *Arbor*, 151 (1995), 35–64.

Kirchner, Helena and Navarro, Carmen. 'Objetivos, métodos y práctica de la arqueología hidráulica', *Archeologia Medievale*, 20 (1993), 121–50.

Lacort Navarro, Pedro J. 'Obras hidráulicas e implantación rural romana en la campiña de Córdoba', *AZA*, I, 359–404.

Ladero Quesada, Miguel Angel, ed. *La incorporación de Granada a la Corona de Castilla.* Granada, Diputación Provincial, 1993.

Ladero Quesada, Miguel Angel and González Jiménez, Manuel. 'La población en la frontera de Gibraltar y el Repartimiento de Vejer (siglos XIII y XIV)', *Historia, Instituciones, Documentos*, 4 (1977), 199–317.

Lagardère, Vincent. *Campagnes et paysans d'Al-Andalus (VIIIe–XVe s.).* Paris, Maisonneuve et Larose, 1993.

Laliena, Carlos and Sénac, Philippe. *Musulmans et Chrétiens dans le Haut Moyen Age: Aux origines de la reconquête aragonaise.* Toulouse, Minerve, 1991.

Lauranson-Rosaz, Christian. *L'Auvergne et ses marges (Velay, Gévaudan) du VIIIe au XIe siècle. La fin du Monde Antique?* Le Puy-en-Velay, Cahiers de la Haute-Loire, 1987.

Lentisco Puche, José Domingo. *La repoblación de Olula del Río (Almería) en el siglo XVI.* Almería, Instituto de Estudios Almerienses, 1991.

Linehan, Peter. *History and Historians of Medieval Spain.* Oxford, Clarendon Press, 1993.

López Andrés, José and Martín-Caro Saura, Faustino. 'Organización, distribución y problemas derivados de la administración del agua en Almería y su vega en los años anteriores a la conquista', *AZA*, II, 1017–32.

López Andrés, Jesús M. 'La intervención de la iglesia de Almería en la administración de las aguas del abasto del común de la ciudad', *AZA*, II, 863–73.

López de Coca Castañer, José E. *La tierra de Málaga a fines del siglo XV.* Granada, Universidad, 1977.

—— *El Reino de Granada en la época de los Reyes Católicos.* 2 vols. Granada, Universidad, 1989.

López Elum, José. *La alquería islámica en Valencia: Estudio arqueológico de Bofilla, siglos XI a XVI.* Valencia, privately printed, 1994.

López Quiroga, Jorge and Rodríguez Lovelle, Mónica. 'Una aproximación arqueológica al problema historiográfico de la "despoblación y repoblación en el valle del Duero", s. VIII–XI', *Anuario de Estudios Medievales*, 21 (1991), 3–9.

Luna Diaz, Juan Andrés. 'La alquería: un modelo socio-económico en la vega de Granada. Aproximación a su estudio', *Chronica Nova*, 16 (1988), 79–100.

—— 'Repoblación y gran propiedad en la región de los Montes de Granada durante el siglo XVI. El cortijo', *Chronica Nova*, 17 (1989), 171–204.

McKeon, Richard. 'The Organization of Sciences and the Relations of Cultures in the Twelfth and Thirteenth Centuries', in J. Murdoch and E. Sylla, eds., *The Cultural Context of Medieval Learning* (Dordrecht, D. Reidel, 1975), 151–92.

Maass, Arthur, and Anderson, Raymond L. . . . *And the Desert Shall Rejoice: Conflict, Growth and Justice in Arid Environments.* Cambridge, MA, MIT Press, 1978.

Magnou-Nortier, Elisabeth. *La société laïque et l'église dans la province ecclésiastique de Narbonne de la fin du VIIIe à la fin du XIe siècle.* Toulouse, Université de Toulouse-Le Mirail, 1974.

—— 'La gestion publique en Neustrie: Les moyens et les hommes (VIIe–IXe siècles)', in H. Atsma, ed., *La Neustrie: Les pays du nord de la Loire de 650 a 850.* 2 vols. (Sigmaningen, Jan Thorbecke Verlag, 1989), I, 271–318.

Malpica Cuello, Antonio. *Turillas, alquería del Alfoz seixitano.* Granada, Universidad, 1984.

—— 'Poblamiento del Reino de Granada. Estructuras nazaríes y modificaciones castellanas', in *Les Illes Orientals d'Al-Andalus* (Palma de Mallorca, Institut d'Estudis Baleàrics, 1987), pp. 375–93.

—— 'Un modelo de ocupación humana del territorio de la Alpujarra: Las ta'a/s de

Sahil y Suhayl a fines de la Edad Media', in *Sierra Nevada y su entorno* (Granada, Casa de Velázquez/Universidad, 1988), pp. 293–315.

—— 'La implantación castellana en la tierra de Salobreña. La alquería de Benardila', *Revista del Centro Histórico de Granada y su Reino*, 2nd epoch, 3 (1989), 199–213.

—— 'De la Granada nazarí al reino de Granada', in *De Al-Andalus a la sociedad feudal* (Barcelona, CSIC, 1990), pp. 119–53.

—— 'Repoblaciones y nueva organización del espacio en zonas costeras granadinas', in Miguel Angel Ladero Quesada, ed., *La incorporación de Granada a la Corona de Castilla* (Granada, Diputación Provincial, 1993), pp. 513–58.

Malpica Cuello, Antonio (ed.) *La cerámica altomedieval en el sur de Al-Andalus*. Granada, Universidad, 1993.

Manzano, Eduardo. 'El regadío en Al-Andalus: Problemas en torno a su estudio', *En la España medieval* (Madrid), 5 (1986), 617–32.

—— 'Beréberes de Al-Andalus: Los actores de una evolución histórica', *Al-Qanṭara*, 11 (1990), 397–428.

—— *La Frontera de al-Andalus en época de los omeyas*. Madrid, CSIC, 1991.

Marín-Guzmán, Roberto. 'The Revolt of ᶜUmar ibn Ḥafṣūn: A Challenge to the Structure of the State (880–928)'. Unpublished doctoral dissertation, University of Texas at Austin, 1994.

Martí, Ramon. 'Hacía una arqueología hidráulica. La génesis del molino feudal en Cataluña', in M. Barceló, ed., *Arqueología medieval* (Barcelona, Crítica, 1988), pp. 165–94.

—— 'Les *insulae* medievals catalans', *Butlletí de la Societat d'Arqueologia Lul·liana*, 44 (1988), 11–23.

—— 'Sistemes hidràulics i poblament en els límits de Catalunya Vella: la unitat hidrològica del riu de Bitlles (Anoia/Alt Penedès)', *IV CAME*, III, 587–93.

Martínez Sanmartín, Luis Pablo. 'Estructura social y cambio tecnológico. Una crítica a los determinismos tecnológico y economicista en la historia de la técnica', *Arbor*, 143 (1992), 103–31.

—— 'El estudio social de los espacios hidráulicos', *Taller d'Història* (Valencia), 1 (1993), 90–3.

—— 'La lluita per l'aigua com a factor de producció. Cap a un model conflictivista d'anàlisi dels sistemes hidràulics valencians', *Afers*, 15 (1993), 27–44.

Martínez Sopeña, Pascual. *La tierra de campos occidental. Poblamiento, poder y comunidad del siglo X al XIII*. Valladolid, Institución Cultural Simancas, 1985.

Matesanz Vera, Pedro. 'La cerámica medieval cristiana en el norte (ss. IX–XIII): Nuevos datos para su estudio', *II CAME*, I, 245–60.

Mateu Bellés, Joan F. 'Assuts i vores fluvials regades al País Valencià', in *Los paisajes de agua. Libro jubilar dedciado al profesor Antonio López Gómez* (Valencia/Alicante, Universidad, 1989), pp. 165–85.

Minguez, José María. 'Ruptura social e implantación del feudalismo en el noroeste penínsular (siglos VIII–X)', *Studia Histórica. Historia Medieval*, 3:2 (1985), 7–32.

Molénat, Jean-Pierre. 'Villes et forteresses musulmanes de la région toledane disparues après l'occupation chrétienne', *Castrum 3*, 215–24.

Molina, Luis. *Una descripción anónima de Al-Andalus*. 2 vols. Madrid, CSIC, 1983.

—— 'Orosio y los geógrafos hispanomusulmanes', *Al-Qanṭara*, 5 (1984), 63–92.

Moreta Velayos, Salustiano. *El monasterio de San Pedro de Cardeña: Historia de un dominio monástico castellano (902–1338)*. Salamanca, Universidad, 1971.

Muñoz Buendia, Antonio and Diaz López, Julián Pablo. 'Continuidad y cambio de la estructura agararia almeriense en la edad moderna: El caso de Pechina', in *Almería entre dos culturas. Siglos XIII al XVI*, 2 vols. (Almería, Instituto de Estudios Almerienses, 1990), II, 731–62.

Mut Calafell, Antoni and Rosselló Bordoy, Guillem, eds. *La remembrança de Nunyo Sanç: Una relació de les seves propietats a la ruralia de Mallorca*. Palma de Mallorca, Museu de Mallorca, 1993.

Navarro, Carmen. 'El ma'gil de Liétor: un sistema de terrazas de origen andalusí en activo', I Congreso de Arqueología Peninsular, Porto, 1993 (in press).

Oliver Asín, Jaime. *Historia del nombre 'Madrid'.* Madrid, CSIC, 1958.

Olmo Enciso, Lauro. 'Ideología y arqueología: los estudios sobre el período visigodo en la primera mitad del siglo XX', in Javer Arce and Ricardo Olmo, eds., *Historiografía de la arqueología y de la historia antigua en España (siglos XVIII–XX)* (Madrid, Ministerio de Cultura, 1991), pp. 157–60.

—— 'El reino visigodo de Toledo y los territorios bizantinos. Datos sobre la heterogeneidad de la península ibérica', *Coloquio Hispano-Italiano de Arqueología Medieval* (Granada, Patronato de la Alhambra, 1992), pp. 185–98.

Orejas Saco del Valle, Almudena, and Sánchez Palencia, F. Javier. 'Obras hidráulicas romanas en la provincia de Toledo', *AZA*, I, 43–67.

Osorio Pérez, María José and Peinado Santaella, Rafael G. 'El Libro de Repartimiento de Montejicar (1527): Comentario y edición', *Revista del Centro de Estudios Históricos de Granada y su Reino,* 2nd epoch, 4 (1990), 71–112.

Parejo Delgado, María Josefa. 'El abastecimiento urbano de Baeza y Ubeda en la baja edad media', *AZA,* II, 813–36.

Pastor, Reyna. *Resistencia y luchas campesinas en la época del crecimiento y consolidación de la formación feudal. Castilla y León, siglos X–XIII.* Madrid, Siglo Veintiuno, 1980.

—— 'Sobre la constitución y consolidación del sistema feudal castellano-leonés de los siglos XI–XII', *Estudi General,* 5–6 (1985–86), 199–210.

Peinado Santaella, Rafael G. *La repoblación de la tierra de Granada: Los montes orientales (1485–1525).* Granada, Universidad, 1989.

Peñarroja Torrejón, Leopoldo. *Cristianos bajo el Islam: los Mozárabes hasta la reconquista de Valencia.* Madrid, Gredos, 1993.

Pérez-Embid Wamba, Javier. *El Cister en León y Castilla.* Junta de León y Castilla, n.p., 1986.

Philippe, Robert. 'L'église et l'énergie pendant le XIe siècle dans les pays d'entre Seine et Loire', *Cahiers de Civilisation Médiévale,* 29 (1984), 107–17.

Pica, Christophe. 'L'Evolution des localités de l'Algarve du XIème au XIIIème siècles', *Cahiers d'Histoire,* 37 (1992), 3–21, esp. pp. 9–16.

Picard, Christophe. 'L'Evolution des localités de l'Algarve du XIème au XIIIème siècles', *Cahiers d'Histoire,* 37 (1992), 3–21.

Pirenne, Jacqueline. *La maîtresse de l'eau en Arabie du Sud antique.* Paris, Institut de France, 1977.

Pocklington, Robert. 'Acequias árabes y pre-árabes en Murcia y Lorca: Aportación toponímica a la historia del regadío', *Xé Col·loqui General de la Societat d'Onomàstica. 1er d'Onomàstica Valenciana* (Valencia, Universidad, 1986), pp. 462–73.

—— 'Toponimia y sistemas de agua en Sharq al-Andalus', in Míkel de Epalza, ed., *Agua y poblamiento musulmán* (Benissa, Ajuntament, 1988), pp. 103–14.

—— *Estudios toponímicos en torno a los orígenes de Murcia.* Murcia, Academia Alfonso X el Sabio, 1990.

Ponce Molina, Pedro. *El espacio agrario de Fondón en el siglo XVI.* Fondón, Ayuntamiento, 1984.

—— *Repartimiento de Dalias/El Ejido.* Almería, Quash/Tierras de Almería, 1984.

—— 'Moriscos y repobladores. El paisaje agrario de Adra en la segunda mitad del siglo XVI', in *Almería entre dos culturas* (Almería, Instituto de Estudios Almerienses, 1990), II, 839–59.

Ponsot, Pierre. 'Les Morisques, la culture irrigué du blé et le problème de la décadence de l'agriculture espagnole au XVIIe siècle', *Mélanges de la Casa de Velázquez,* 7 (1971), 237–62.

Portela, Ermelindo, and Pallares, María Carmen. 'De la villa altomedieval a la fortaleza del siglo XV. Fuentes escritas y arqueología en Galicia', in *Coloquio Hispano-Italiano de Arqueología Medieval* (Granada, Patronato de la Alhambra, 1992), pp. 215–21.

Poveda Sánchez, Angel. 'Introducción al estudio de la toponimia árabe-musulmana de Mayurqa según la documentación de los Archivos de la Ciutat de Mallorca (1232–1278)', *Awraq,* 3 (1980), 75–102.

—— 'Sobre los distritos, las explotaciones y la toponimia clánica de Yabisa (Eivissa)', *Sharq Al-Andalus,* 1 (1984), 109–15.

—— 'Toponimia àrabo-berber i espai social a les Illes Orientals d'Al-Andalus'. Unpublished doctoral dissertation, Universitat Autònoma de Barcelona. Bellaterra, 1987.

Retuerce Velasco, Manuel, and Canto García, Alberto. 'Apuntes sobre la cerámica emiral a partir de dos piezas fechadas por monedas', *II CAME*, III, 93–104.

Retuerce Velasco, Manuel, and Zozaya, Juan. 'Variantes geográficas de la cerámica omeya andalusí: los temas decorativos', in *La ceramica medievale nel Mediterraneo occidentale* (Florence, Edizioni all'Insegna del Giglio, 1986), 69–128.

Riu Riu, Manuel. 'Poblados mozárabes de Al-Andalus. Hipótesis para su estudio: El ejemplo de Busquistar', *Cuadernos de Estudios Medievales*, 2–3 (1974–75), 3–35.

—— 'Aportación de la arqueología al estudio de los mozárabes de Al-Andalus', in Riu Riu *et al.*, *Tres estudios de historia medieval andaluza*, 2nd ed. (Córdoba, Caja de Ahorros, 1982), pp. 85–112.

—— 'L'Aportació de l'arqueologia a l'estudi de la formació i expansió del feudalisme català', *Estudi General*, 5–6 (1985–86), 27–45.

—— *L'Arqueologia medieval a Catalunya.* Barcelona, Llibres de la Frontera,1989.

—— 'Talleres y hornos de alfareros de cerámica gris en Cataluña', in *Fours de potiers et 'testares' médiévaux en Méditérranée occidentale* (Madrid, Casa de Velázquez, 1990), pp. 105–15.

Rodríguez López, Juana M. and Cara Barronuevo, Lorenzo. 'Aproximación al conocimiento de la historia agrícola de la Alpujarra oriental (Almería). Epocas antigua y medieval', *AZA*, I, 441–66.

Rosselló Bordoy, Guillermo. *El nombre de las cosas en Al-Andalus: Una propuesta de terminología cerámica.* Palma de Mallorca, Museu de Mallorca, 1991.

—— 'Precisiones sobre terminología cerámica andalusí', *Coloquio Hispano-Italiano de Arqueología Medieval* (Granada, Patronato de la Alhambra, 1992), pp. 253–62.

—— 'Las cerámicas de primera época: Algunas observaciones metodológicas', in Antonio Malpica Cuello, ed., *La cerámica altomedieval en el sur de Al-Andalus* (Granada, Universidad, 1993), pp. 13–25.

Rubiera, María Jesús. 'Rafals y reales; Ravals y arrabales; Reals y reales', *Sharq al-Andalus*, 1 (1984), 117–22.

—— 'Toponimia arábigo-valenciana: Falsos antropónimos beréberes', in *Miscel·lània Sanchis Guarner* (Valencia, Universidad, 1984), pp. 317–20.

Rubiera, María Jesús, and Epalza, Míkel de. *Xàtiva musulmana (segles VIII–XIII).* Xàtiva, Ajuntament, 1987.

—— *El noms àrabs de Benidorm i la seua comarca.* Alicante, Universitat, 1985.

Rucquoi, Adeline. 'Molinos et *aceñas* au coeur de la Castille septentrionale (XIe–XVe siècles)', in *Les Espagnes médiévales. Aspects économiques et sociaux* (Nice, Faculté des Lettres, 1983), pp. 107–22.

Ruiz, Teófilo. *Sociedad y poder real en Castilla.* Barcelona, Ariel, 1981.

Sáenz de Santa María, Antonio *Molinos hidráulicos en el Valle Alto del Ebro (s. IX–XV).* Vitoria, Diputación Foral de Alava, 1985.

Sā'id al-Andalusī. *Science in the Medieval World: Book of the Categories of Nations*, ed. S. I. Salem and A. Kumar, Austin, University of Texas Press, 1991.

Sales Martínez, Vicente. 'La cuestión del extremal en el regadío de la Real Acequia de Montcada', *Cuadernos de Geografía* (Valencia), 44 (1988), 221–34.

Salrach, Josep M. 'Del estado romano a los reinos germánicos. En torno a las bases materiales del poder del estado en la Antigüedad tardia y la alta edad media', in *De la Antigüedad al Medioevo* (Madrid, Fundación Sánchez-Albornoz, 1993), pp. 95–142.

Salvatierra Cuenca, Vicente. *Cien años de arqueología medieval. Perspectivas desde la periferia: Jaén.* Granada, Universidad, 1990.

Samsó, Julio. 'La tradición clásica en los calendarios agrícolas hispanoárabes y norteafricanos', in *Second International Congress of Studies on Cultures of the Western Mediterranean* (Barcelona, n.p., 1978), pp. 177–86.

—— 'The Early Development of Astrology in Al-Andalus', *Journal for the History of Arabic Science*, 3 (1979), 228–43.

—— 'La primitiva versión árabe del Libro de las Cruces', in Juan Vernet, ed., *Nuevos estudios sobre astronomía española en el siglo de Alfonso X* (Barcelona, Institución Milá i Fontanals, 1983), pp. 149–61.

—— 'En torno a los métodos de cálculo utilizados por los astrólogos andalusíes a fines del s. VIII y principios del IX: Algunas hipótesis de trabajo', *Actas de las II Jornadas de Cultura Arabe e Islámica (1980)* (Madrid, Instituto Hispano-Arabe de Cultura, 1985), pp. 509–22.

—— 'Nota sobre la biografía del rey Sisebuto en un texto árabe anónimo', *Serta gratulatoria in honorem Juan Régulo,* I Filología (La Laguna, Universidad, 1985), pp. 639–42.

—— 'Astrology, Pre-Islamic Spain and the Conquest of Al-Andalus', *Revista del Instituto Egipcio de Estudios Islámicos en Madrid,* 23 (1985–86), 79–94.

—— *Las ciencias de los antiguos en Al-Andalus.* Madrid, Mapfre, 1992.

Sánchez-Albornoz, Claudio. 'Los libertos en el reino astur-leonés', *Instituciones medievales españolas* (Mexico, UNAM, 1965), pp. 317–51.

—— *Despoblación y repoblación del valle del Duero.* Buenos Aires, Instituto de Historia de España, 1966.

—— 'Pequeños propietarios libres en el reino asturleonés. Su realidad histórica', in *Investigaciones y documentos sobre las instituciones hispanas.* Santiago, Editorial Jurídica de Chile, 1970, pp. 178–210.

—— 'La repoblación del reino asturleonés. Proceso, dinámica y proyecciones', *Cuadernos de Historia de España,* 53–4 (1971), 236–459.

Sanz Fuentes, María Josefa. 'Repartimiento de Ecija', in *Historia, Instituciones, Documentos,* 3 (1976), 533–51.

Scales, Peter C. *The Fall of the Caliphate of Cordoba: Berbers and Andalusis in Conflict.* Leiden, E. J. Brill, 1994.

Segura Graiño, Cristina. 'Los Repartimientos medievales andaluces: Estado de la cuestión', *Anuario de Estudios Medievales,* 12 (1982), 625–39.

—— *El Libro del Repartimiento de Almería.* Madrid, Universidad Complutense, 1982.

Selma Castell, Sergi. 'La integración de los molinos en un sistema hidráulico: La alquería de Artana (Serra d'Espadà, Castelló)', *AZA,* II, 713–36.

—— 'El molí hidràulic de farina i l'organització de l'espai rural andalusí. Dos exemples d'estudi arqueològic espacial a la Serra d'Espadà (Castelló)', *Mélanges de la Casa de Velázquez,* 27 (1991), 65–100.

—— 'Notes sobre la formació d'uns primers monopolis feudals a la Vall d'Albaida', *Alba* (Ontinyent), 7 (1992), 35–8.

—— 'Toponimia tribal i clànica d'origen berber al nord de Sharq al-Andalus (Recull i noves prepostes)', *Estudis Castellonencs,* 5 (1992–93), 459–66.

—— 'Molins i rodes: entorn d'una discusió desafortunada', *Afers* (Valencia), 15 (1993), 11–26.

—— 'Conquesta feudal i creació de monopolis de renda al País Valencià', *Boletín de la Sociedad Castellonense de Cultura,* 69 (1993), 333–55.

—— *Els molins d'aigua medievals a Sharq al-Andalus.* Onda, Ajuntament, 1993.

—— 'Evolució des de l'epoca andalusí de l'espai agrari irrigat a la Vall de Veo (Serra d'Espadà, Castelló)', *IV CAME,* III, 567–74.

Sénac, Philippe. 'Notes sur les *ḥuṣūn* de Lérida', *Mélanges de la Casa de Velázquez,* 24 (1988), 53–69.

—— 'Poblamiento, habitats rurales y sociedad en la Marca Superior de Al-Andalus', *Aragón en la Edad Media,* 9 (1991), 389–401.

Sénac, Philippe, and Esco, Carlos, 'Le peuplement musulman dans le district de Huesca (VIIIe–XIIIe siècles)', in *La Marche Supérieure d'Al-Andalus et l'Occident Chrétien* (Madrid, Casa de Velázquez, 1991), pp. 51–65.

Shaw, Brent D. 'Lamasba: An Ancient Irrigation Community', *Antiquités Africaines* (1982), 61–103.

Soto i Company, Ricard, ed. *Còdex català del Llibre del Repartiment de Mallorca.* Palma de Mallorca, J. J. de Olañeta, 1984.

—— 'Repartiment i Repartiments; l'ordenació d'un espai de colonització feudal a

Mallorca del segle XIII', in *De Al-Andalus a la sociedad feudal: Los Repartimientos bajomedievales* (Barcelona, CSIC, 1990), pp. 1–51.
Torres Fontes, Juan. *Repartimiento de Murcia.* Madrid, CSIC, 1960.
—— 'Los Repartimientos murcianos del siglo XIII', in *De Al-Andalus a la sociedad feudal: Los repartimientos bajomedievales* (Barcelona, CSIC, 1990), pp. 71–94.
—— *Repartimiento de Lorca.* 2nd ed. Murcia, Academia Alfonso X el Sabio, 1994.
Torres Fontes, Juan, ed. *Repartimiento de Orihuela.* Murcia, Academia Alfonso X el Sabio, 1988.
Torró, Josep. 'El problema del hábitat fortificado en el sur del reino de Valencia después de la segunda revuelta mudéjar (1276–1304)', *Anales de la Universidad de Alicante. Historia Medieval,* 7 (1988–89), 53–81.
—— 'Sobre ordenament feudal del territori i trasbalsaments del poblament mudèjar. La *Montanea Valencie* (1286–1291)', *Afers,* 7 (1988–89), 95–124.
—— *Poblament i espai rural: Transformacions històriques.* Valencia, Institució Valenciana d'Estudis i Investigació, 1990.
—— *Alcoi. La formació d'un espai feudal (de 1245 a 1305).* Valencia, Diputació Provincial, 1992.
Toubert, Pierre. *Castillos, señores y campesinos en la Italia medieval.* Barcelona, Crítica, 1990.
Valdeón Baruque, Julio. 'Señores y campesinos en la Castilla medieval', in *El pasado histórico de Castilla y León,* Vol. I. *Edad Media* (Burgos, Junta de Castilla y León, 1983), pp. 59–86.
Vallvé, Joaquín. *Nuevas ideas sobre la conquista árabe de España: Toponimia y onomástica.* Madrid, Real Academia de la Historia, 1989.
Vernet, Juan. 'Los médicos andaluces en el "Libro de la Generaciones de Médicos", de Ibn Juljul', in *Estudios sobre Historia de la Ciencia Medieval* (Barcelona, Universidad, 1979), pp. 469–86.
Viguera Molins, María Jesús. 'En torno a las fuentes jurídicas de al-Andalus', in *Actes, Congrès 'La Civilisation d'al-Andalus dans le temps et dans l'espace'* (Mohammedia, Université Hassan II, 1992–93), pp. 71–8.
—— 'Sobre Mozárabes', in *Proyección histórica de España en sus tres culturas: Castilla y León, América y el Mediterráneo,* vol. III, *Arabe, Hebreo e historia de la medicina* (Valladolid, Junta de León y Castilla, 1993), pp. 205–16.
Vilches Vilches, Carlos. *La Alhambra de Leopoldo Torres Balbás.* Granada, Comares, 1988.
Villar García, Luis Miguel. *La Extremadura castellano-leonesa: Guerreros, clérigos y campesinos (711–1252).* Valladolid, Junta de Castilla y León, 1986.
Vincent, Bernard. 'La société chrétienne almeriense et les systemes hydrauliques. Quelques propositions de travail', in *AZA,* I, xciii–cix.
Virella i Bloda, Albert. 'Els molins d'aigua en l'alta medievalitat a ponent del Llobregat', *Miscel·lània Penedesenca,* 6 (1983), 249–71.
Watson, Andrew M. *Agricultural Innovation in the Early Islamic World.* Cambridge, Cambridge University Press, 1983.
Weissing, Franz and Ostrom, Elinor. 'Irrigation Institutions and the Games Irrigators Play: Rule Enforcement without Guards', in Reinhard Selton, ed., *Game Equilibrium Models, II: Methods, Morals, and Markets* (Berlin, Springer-Verlag, 1991), pp. 188–262.
Wickham, Chris. 'The Other Transition: From the Ancient World to Feudalism', *Past and Present,* 103 (May 1984), 3–36.
—— 'L'incastellamento ed i suoi destini, undici anni dopo el *Latium* de P. Toubert', *Castrum* 2, pp. 411–20.
—— *The Mountains and the City: the Tuscan Apennines in the Early Middle Ages.* Oxford, Clarendon Press, 1988.
—— 'L'Italia e l'alto medioevo', *Archeologia Medievale,* 15 (1988), 105–25.
Wittfogel, Karl A. *Oriental Despotism.* New Haven, Yale University Press, 1957.
Yelo Templado, Antonio, *et al.* 'Aportación al estudio del poblamiento y los regadíos de época romana en la cabecera del valle del Segura, fuentes documentales y arqueológicas', *Antigüedad y cristianismo* (Murcia), 5 (1988), 599–611.
Yzquierdo y Pino, Pablo. 'Fonts documentals i ceràmica andalusina. Alguns suggeriments',

in *Segones Jornades de Joves Historiadors i Historiadores* (Barcelona, mimeo, 1988), n.p.

Zozaya, Juan. 'The Islamic Consolidation in Al-Andalus (8th–10th Centuries): An Archaeological Perspective'. Typescript.

—— 'Importaciones casuales en Al-Andalus: Las vías de comercio'. Typescript.

Index